Also by Alan Lightman

Probable Impossibilities

Three Flames

In Praise of Wasting Time

Searching for Stars on an Island in Maine

Screening Room

The Accidental Universe

Mr g

Song of Two Worlds

Ghost

The Discoveries

A Sense of the Mysterious

Reunion

The Diagnosis

Dance for Two

Good Benito

Einstein's Dreams

A Modern Day Yankee in a Connecticut Court

Time Travel and Papa Joe's Pipe

The Transcendent Brain

The Transcendent Brain

Spirituality in the Age of Science

Alan Lightman

Pantheon Books, New York

 Published in the United States by Pantheon Books, a division of Penguin Random House LLC, New York, and distributed in Canada by Penguin Random House Canada Limited, Toronto.

Pantheon Books and colophon are registered trademarks of Penguin Random House LLC.

Grateful acknowledgment is made for permission to reprint "Sleeping in the Forest," by Mary Oliver, copyright © 1978 by Mary Oliver. Reprinted by permission of the Charlotte Sheedy Literary Agency and Bill Reichblum.

Library of Congress Cataloging-in-Publication Data
Names: Lightman, Alan P., [date] author.
Title: The transcendent brain : spirituality in the age of science / Alan Lightman.
Description: First edition. New York : Pantheon Books, 2023. Includes bibliographical references.
Identifiers: LCCN 2022010019 | ISBN 9780593317419 (hardcover) | ISBN 9780593317426 (ebook)
Subjects: LCSH: Religion and science. Spirituality.
Classification: LCC BL240.3 .L54 2023 | DDC 215--dc23 /eng20221028
LC record available at https://lccn.loc.gov/2022010019

www.pantheonbooks.com

Jacket photograph by Junichiro Tokiyoshi / EyeEm / Getty Images
Jacket design by Mark Abrams

Printed in the United States of America
First Edition
9 8 7 6 5 4 3 2

When Figures show their royal Front—
And Mists—are carved away,
Behold the Atom—I preferred—
To all the lists of Clay!

—*Emily Dickinson*

Contents

The Transcendent Brain

Introduction

For many years, a family of ospreys lived near our house on a small island in Maine. Each season, my wife and I observed their rituals and habits. In mid-April, the parents would arrive at the nest, having spent the winter in South America, and lay eggs. In late May or early June, the eggs hatched. As the father dutifully brought fish dinners to the nest each day, the babies would grow bigger and bigger and in mid-August were large enough to make their first flight. Throughout the season, my wife and I recorded all of these comings and goings. We noted the number of chicks each year. We observed when the chicks first began flapping their wings, in early August, a couple of weeks before having the strength to become airborne and leave the nest for the first time. We memorized the different chirps the parents made for danger, for hunger, for the arrival of food. After several years of cataloging such data, we felt that we knew these ospreys pretty well.

Then, one late August afternoon, the two juvenile ospreys of that season took flight for the first time as I stood observing them from my second-floor circular deck. All summer long, they had watched me on that deck as I watched them. The circular deck was about nest high, so to the fledgling birds it must have looked like I was in my nest just as they were in theirs. On this particular afternoon, their maiden flight, they did a wide half-mile loop out over the ocean and then headed straight at me with tremendous speed. A juvenile osprey, although slightly smaller than a full-grown adult, is still a large bird, with powerful and sharp talons. My immediate impulse was to run for cover, since the birds could have ripped my face off. But something held me to my ground. When they were within fifteen or twenty feet of me, the two birds suddenly veered upward and away. But before that dazzling and frightening vertical climb, for about half a second we made eye contact. Words cannot convey what was exchanged between us in that instant. It was a look of connectedness, of mutual respect, of recognition that we shared the same land. It was a look that said, as clear as spoken words, "We are brothers in this place." After the two young ospreys were gone, I found that I was shaking, and in tears. To this day, I don't understand what happened in that half second. But it was a profound connection to nature. And a feeling of being part of something much larger than myself.

I'm a scientist and have always had a scientific view of the world—by which I mean that the universe is made of material stuff, and only material stuff, and that stuff is governed by a small number of fundamental laws. Every phenomenon has a cause, which originates in the physical universe. I'm a materialist. Not in the sense of seeking happiness in cars and nice clothes, but in the literal sense of the word: the belief that everything is made out of atoms and molecules, and nothing more. Yet, I have transcendent experiences. I communed with two ospreys that summer in Maine. I have feelings of being part of things larger than myself. I have a sense of connection to other people and to the world of living things, even to the stars. I have a sense of beauty. I have experiences of awe. And I've had transporting creative moments. Of course, all of us have had similar feelings and moments. While these experiences are not exactly the same, they have sufficient similarity that I'll gather them together under the heading of "spirituality."

I will call myself a *spiritual materialist*. In 1973, the late Tibetan Buddhist master Chögyam Trungpa coined that phrase to mean someone who has the (false) belief that certain temporary states of mind bring relief from suffering. By "temporary states of mind," he was probably thinking of the pleasures of cars and nice clothes and maybe romance. My meaning is different. I believe that the spiritual experiences we have can arise from atoms and molecules. At the same time, some of these experiences, and certainly their very personal and subjective nature, cannot be fully understood in terms of atoms and molecules. I believe in the laws of chemistry and biology and physics—in fact, as a scientist I much admire those laws—but I don't think they capture, or can capture, the first-person experience of making eye contact with wild animals and similar transcendent moments. Some human experiences are simply not reducible to zeros and ones.

I'll assume that the feelings I've described occur in the brain, possibly augmented by the complete nervous system. In the view of modern biology, all mental sensations are rooted in the material neurons of the nervous system and the electrical and chemical interactions between them. Considering that assumption, a more specific and perhaps blunt way to phrase our central question is: How can the material neurons in the human nervous system give rise to feelings of spirituality?

In recent years, scientists and others have come to recognize events and processes we call "emergent phenomena"—behaviors of complex systems that are not evident in their individual parts. A good example is the way that certain groups of fireflies synchronize their flashing. When a bunch of such fireflies first find themselves together in a field on a summer night, each insect in the group flashes at different random times and at different rates, like blinking Christmas tree lights. But after a minute or so, even without a boss firefly giving orders, all the fireflies have adjusted their inner bodies so that they flash on and off in total synchrony. Such collective behavior cannot be understood by the analysis of a single firefly. Similarly, our brains, composed of 100 billion neurons/fireflies, exhibit all kinds of spectacular behavior that cannot be explained or predicted in terms of individual neurons. The concept of emergent phenomena offers a possible understanding of how a purely materialist world can be compatible with complex human experiences.

Even more fundamental than spirituality is the bedrock experience we call consciousness—the sense of being, of self-awareness, of "I-ness," of existing as a distinct entity able to feel and think. How can the material neurons of the nervous system give rise to the sensation we call consciousness?

Although the deeper question of consciousness must be considered in any discussion of the mind, I'm more interested in the question of spirituality, for several reasons. Consciousness, while we all experience it, is extremely subtle, hard to define, and remains elusive to neurobiologists, philosophers, and psychologists alike. Perhaps with consciousness we are asking the wrong question altogether. The distinguished neuroscientist Robert Desimone of MIT told me that the mystery of consciousness is "overrated," that it is just a vague name we give to the sensation of all the electrical and chemical activity of our neurons. I am not fully satisfied with Professor Desimone's statement. While I completely support the idea that consciousness is rooted in the material nervous system, I don't think that the mystery of consciousness is exaggerated. The sensations of

"I-ness" and immediate presence in the world are unlike all other experiences and equal to the greatest mysteries in science.

If we take consciousness as a given, the various feelings of spirituality mentioned earlier are somewhat specific and definable. We can explore the nature of those feelings, their origins, and their possible evolutionary benefit. Of course, we cannot really take consciousness as a given, since all human experiences, including spirituality, are based on consciousness. So, part of my investigation will be to travel again the well-trod road of attempting to understand consciousness in terms of the physical brain and nervous system. While doing so, I don't wish to diminish in any way the magnificent and utterly unique experiences of consciousness and spirituality that follow naturally from it.

But perhaps more importantly, an investigation of spiritual materialism, beyond consciousness in general, takes us in other directions. It provides a framework for understanding spirituality without reference to an all-powerful, intentional, and supernatural Being (God). Such an exploration relates to intellectual movements like secular humanism—the idea that humans can live a moral and self-fulfilled life without belief in God—while at the same time recognizing the importance and validity of spiritual experiences. Such an investigation also counters the notion that science and spirituality are mutually exclusive.

To be sure, this investigation is nuanced and easily

misinterpreted. In an article titled "Does God Exist?" I published in *Salon* a decade ago, I argued that we have experiences and beliefs that are beyond the reach of science. In particular, we have beliefs that cannot be proved and must be accepted as a matter of faith, such as the belief that the universe was created with a purpose, or the belief that the universe is always lawful. At the same time, I identified myself as a card-carrying scientist. Soon after my article was published, some people from the neo-atheist camp attacked me as being an apologist for religion and for excusing the "fuzzy" and "noncritical" thinking of "believers." In response, I ask: Was Abraham Lincoln a fuzzy thinker? Was Mahatma Gandhi a fuzzy thinker? My aim is not to prove or disprove the existence of God—in my view probably a futile task for both religion and science—but to acknowledge the broader spiritual experiences we have as human beings, outside of a religious context, and to try to understand them as a scientist. My journey will not be filled with certainties or black-and-white pronouncements.

If one starts by assuming the existence of God, then the explanation of spirituality is pretty straightforward. Namely, the origins and perhaps even meaning of spirituality can be assigned to God. God grants us an immortal soul, which connects us to the cosmos. Our sense of moral behavior and goodness and beauty can be traced to God, as discussed by Saint Augustine and others. If we assume the existence of God, then the problem of spirituality finds a ready solution, and many people prefer this explanation. On the other hand, if we don't assume such a Being, then accounting for spirituality becomes more challenging and certainly more in line with a scientific view of the world. This more difficult path is the one I've chosen to follow.

In the first two chapters, I'll explore a bit of the history of the subject, starting with the nonmaterialist view of the world and then moving to the materialist view. Of course, the prime example of nonmaterialism, as understood by most religions, is God. However, the nonmaterialist view concerns far more than a belief in God. It encompasses an entire ethereal world, which may include the immortal soul, Heaven and Hell, a nonmaterial mind separate from the physical body, ghosts, and other such things. I want to understand how this conception originated and the motivations behind it. In my view, belief in an ethereal world, extending from the grave goods and burial rituals of Neanderthals to the men and women of today, represents something deep in our psychology and is not unrelated to our feelings of spirituality.

These first two chapters are not meant to be exhaustive in any way. My intention here is to address some of the high points of these histories, amplified by my own commentary, and to provide some background for the later discussion of brains, consciousness, and spirituality. Chapter 3 grapples with the brain as a physical object and explores the forever puzzling question of how consciousness can arise from a material brain and nervous system. Neuroscientists, philosophers, and psychologists have engaged in a great deal of research on these topics. Much is known. Much remains unknown. I'll try to give a picture of some of the main arguments and conclusions of this research. Chapter 4 will take consciousness as a given and suggest that spirituality naturally emerges from brains/minds with a high level of consciousness and intelligence subject to the forces of natural selection. Each of these chapters will be organized around a leading figure: for chapter 1, Moses Mendelssohn, whom I consider to have given some of the most rational arguments for the soul; for chapter 2, the ancient Roman poet and philosopher Lucretius, who was one of the earliest and most eloquent materialists; for chapter 3, the contemporary neuroscientist Christof Koch, a leader in the material understanding of consciousness; for chapter 4, contemporary social psychologist Cynthia Frantz, who has studied the psychological and social basis for our connection to nature and things larger than ourselves. In the last chapter, I'll return to the overarching idea of spiritual materialism and its importance in the world today. As our nation, and our world, have become more polarized in recent years, the dialogue between science and spirituality has assumed greater and greater importance. Science and religion/spirituality are the two most powerful forces that have shaped human civilization. Neither is going away. Both are part of being human. We are experimenters, and we are also experiencers.

1

The *Ka* and the *Ba*

A Brief History of the Soul, the Nonmaterial, and the Mind-Body Duality

The man sits at the table, leans toward a friend in the opposite chair. One hand rests on his knee, the other lightly cradles his chin with its short scraggly beard. He wears a red jacket, dark pants, silver-buckled shoes, a white shirt with ruffled cuffs. While his friend reaches out with a smile, our man seems lost in some deep inner realm, as if brooding over the vast cosmos of earthly existence and what might come after. His face would be recognized by many in eighteenth-century Europe, from numerous portraits rendered on porcelain teacups, vases and pendants, busts, paintings. His name is Moses Mendelssohn.

This particular painting with the red jacket depicts a meeting between Mendelssohn and two other thinkers: the German writer and philosopher Gotthold Ephraim Lessing and the Swiss poet and theologian Johann Kaspar Lavater. The latter once described Mendelssohn as "a companionable, brilliant soul, with piercing eyes, the body of an Aesop—a man of keen insight, exquisite taste and wide erudition . . . frank and open-hearted."

Let's describe the scene a bit more. Judging from Mendelssohn's visage, he is about fifty years old, making the year about 1779. A chessboard rests on the table. Above it hangs a brass fixture, whose top section is a chandelier and lower part an oil lamp used for the Sabbath and other Jewish holidays. Mendelssohn is the most famous Jew of his generation. Although deeply religious, he has crossed the border from Jew to Gentile. Breaking from a prescribed life of studying the Talmud and Torah in Hebrew, Mendelssohn has mastered the German language, more adeptly than the Prussian king Frederick the Great, and writes his many philosophical works in that tongue. Against the back wall of the room is a shelf filled with books. A wood floor. A beamed ceiling. A richly embroidered green cloth on the table. A woman enters the room carrying a tray with teacups. This is Mendelssohn's home, on 68 Spandau Street in Berlin. It is a prosperous house. After beginning life as the son of a poor Torah scribe and living for years as a lowly clerk in a silk factory, Mendelssohn has become part owner of the factory.

I start with Mendelssohn because no other philosopher or theologian in the history of recorded thought has argued so rationally for the existence of the soul, the prime example, after God, of the nonmaterial. Aristotle claimed that the soul could not exist without a body.

Augustine attributed all aspects of the soul to the perfection of God, Augustine's starting point in all things. Maimonides assumed the existence of the soul, which would become immortal for the virtuous (but not for the sinners). Mendelssohn made none of these assumptions. Coming of age after the scientific revolution of Galileo and Newton, Mendelssohn started from scratch. He constructed logical arguments for the existence of the soul and its immortality. He thought like a scientist as well as a philosopher. In 1763, he won the prize offered by the Prussian Royal Academy of Science for an essay on the application of mathematical proofs to metaphysics, beating out such people as Immanuel Kant. In his salon, a portrait of Isaac Newton hung next to the portraits of the Greek philosophers.

Mendelssohn was a polymath. As a boy, he studied astronomy, mathematics, philosophy. He wrote poetry. He played the piano (studying with a student of J. S. Bach). At the age of sixteen, he began learning Latin, so that he could read Cicero and a Latin version of John Locke's *Essay Concerning Human Understanding.* Aaron Gumperz, the first Prussian Jew to become a medical doctor, taught Mendelssohn French and English. In his twenties, Mendelssohn joined the German writer and bookseller Christoph Friedrich Nicolai to publish the literary journals *Bibliothek* and *Literaturbriefe*. Not content with five languages under his belt, Mendelssohn then learned Greek, so that he could read Homer and Plato in the original.

In 1767, Mendelssohn wrote his masterwork, *Phädon, or On the Immortality of the Soul,* a reconception of Plato's famous *Phaedo*. In doing so, Mendelssohn wanted to do for the modern European world what Plato had done for the ancient Greek world—describe the necessity and nature of the soul. "I . . . tried to adapt the metaphysical proofs to the taste of our time," Mendelssohn modestly wrote in the preface to his book. But he did more than adapt. He presented new arguments. He reasoned that while the body and all experiences of the body are composed of parts, to arrive at *meaning* there must be a thinking thing outside of the parts to integrate and lead their individual sensations, just as a conductor is needed to lead a symphony orchestra.

Furthermore, this thinking thing beyond the body must be a whole. If it were composed of parts, then there would need to be another thing outside of it, which composed and integrated its parts, and so on, ad infinitum. "There is, therefore . . . at least one single substance, which is not extended, not compound, but is simple, has a power of intellect, and unites all our concepts, desires, and inclinations in itself. What hinders us from calling this substance our soul?" And, the Jewish scholar argued, the soul must be immortal, because nature always proceeds in gradual steps.

Nothing in the natural world leaps from existence to nothingness.

Mendelssohn was a firm believer in God and mentions God frequently in his *Phädon*. But unlike most of his predecessors, many of his arguments for the existence and nature of the immaterial soul did not depend on the existence of God.

Phädon was an immediate success. The first edition sold out in four months. It was translated into Dutch, French, Italian, Danish, Russian, and Hebrew. It portrayed man as a noble being, aspiring to truth and perfection. Perhaps more importantly, it gave eighteenth-century Europe a rational argument for the existence and immortality of the soul, at a time when materialistic views were widespread in an extension of the mechanical world of the scientific revolution. Mendelssohn fought fire with fire. The scientific worldview of Newton and others had reduced the cosmos to a system of levers and pulleys. Mendelssohn used that same logic of scientific reasoning to argue for a nonmaterial essence, a soul, something far beyond levers and pulleys.

Mendelssohn was a shining star of the Enlightenment, joining the constellation of Leibniz and Kant and Goethe. He was called "the German Socrates." He never attended university.

I'm connected to Mendelssohn—through the piano. I've got an upright myself, a Baldwin Acrosonic, and have recently been playing the "Venetian Boat Song," composed by Mendelssohn's grandson, Felix. (Like his grandfather, Felix could speak several languages.) But there's more. My own piano teacher was a grandstudent of the composer and piano virtuoso Franz Liszt. Franz and Felix were fierce rivals. (Felix once said of his competitor, "Liszt has many fingers but few brains.")

The "Venetian Boat Song" is one of forty-nine lovely pieces in a collection called *Songs Without Words* (*Lieder ohne Worte*). It has a sadness to it, a longing. I associate those musical feelings with the grandfather Moses. I believe that part of what impelled him in his *Phädon,* aside from the many rational arguments, was a very personal desire for immortality, especially for his family. Two of his children, Sarah and Chaim, had died at a very young age. Could death be the end of existence? All of us ask that question. I think that in *Phädon,* Mendelssohn might have been trying to soothe his family's grief and give them hope, just as in the *Phaedo,* hours before drinking the poisonous hemlock—the death sentence imposed on him for corrupting the youth with his philosophy—Socrates gives his students an argument for the immortality of the soul to relieve their sadness at his impending death.

I feel a kinship with Mendelssohn not only through the piano, but also through our mutual appreciation of science and its thinking. If I could sit at the table in that painting, I'd ask him some questions. I'm sure there was more underneath that glittering intellectual façade than his belief in God. In fact, Mendelssohn had an almost deist view of God. ("[God] does as few miracles as possible," he wrote.) There was more than his personal loss. Even more than his rational mind at work. On purely logical grounds, there's almost certainly a fatal flaw in his principal argument for the soul: that a thing of many parts, like the body, requires something outside of itself to gather up the pieces and make harmony and order. It's a reasonable argument. However, the science of the last century has shown how a system of many parts can create order even within itself, in a process known as emergence, which I referred to in the introduction. The magnificent dirt cathedrals formed by colonies of termites, the patterns of snowflakes, the intricate and highly functional folding arrangements of proteins all show that an external organizing force is not required to produce order and harmony from mindless parts.

I'd enjoy telling Mendelssohn about these ideas in modern science and getting his reaction. Perhaps he might refute me. Or perhaps he might come up with new arguments. But I think that all such arguments are doomed. In my view, the existence of the soul, like the existence of God, cannot be proved by any rational argument. (To put it another way, how could we know for sure that some phenomenon attributed to God could not be explained by a nontheist cause?) Believers in the soul, or God, must accept such beliefs as a matter of faith. Still, I admire Mendelssohn's reasoning. I want to understand the various forces shaping his thought, forces that have endured for thousands of years in our attempt to find meaning and comfort in this strange cosmos we find ourselves in. I want to understand the how and the why of the soul—and indeed all nonmaterial things. Most importantly, I think that belief in the soul, shared by Mendelssohn and other philosophers and theologians, has some of the same psychological and evolutionary underpinnings as other feelings I have associated with spirituality.

Belief in the soul has an ancient history. Its oldest mention, perhaps, can be found in the hieroglyphics carved on the walls of the burial chamber of Unis, pharaoh in the Fifth Dynasty of the Old Kingdom of Egypt, dating back to about 2315 BC:

> Ho, Unis! You have not gone away dead; you have gone away alive . . .
> Dispatches of your ka have come for you, dispatches of your father have
> come for you, dispatches of the Sun have come for you . . . You shall become
> clean in the cool waters of the stars and board the sunboat on cords of metal . . .
> Humanity will cry out to you once the Imperishable Stars have raised you aloft.

The purpose of such incantations was to help the deceased unite with his soul in an afterlife. The ancient Egyptians believed that each human being was composed of three parts: the material body; a nonmaterial element called the *ka,* which was the universal life force and which returned to the gods after death; and the nonmaterial *ba,* which encompassed the unique personality of the individual. At death, the *ka* and *ba* departed from the body. To become an eternal spirit in the afterlife, the *ba* had to be reunited with its *ka*. Both the *ka* and the *ba* were souls.

The idea of multiple kinds of souls can be found in many later conceptions of the spiritual. The Chinese, the Hindus, the Inuits, the Jainists, the Shamanists, the Tibetans all have notions of dual or multiple souls. Mendelssohn has a similar idea. He divides all of existence into three levels: "The first level thinks, but cannot be thought of by others [the universal soul]: this is the only one whose perfection surpasses all finite concepts. The created minds and souls make up the second level [the personal souls]: These think and can be thought of by others. The corporeal world is the last level, which can be thought of only by others, but cannot think." According to Mendelssohn, the purpose of the personal soul is to find ultimate truth, perfection, and wisdom. If we associate perfection with the universal soul, then the personal soul is striving to merge with the universal soul, as the *ba* with its *ka*.

In almost all cultures, the soul has been associated with some kind of life force that distinguishes human beings from rocks. In other words, the soul is a distinctive feature of living things. The Greek and Latin words that are often translated as "soul," *psyche* and *animus,* both refer to life.

The soul is always nonmaterial, often but not always invisible, usually eternal, and usually perfect, in contrast to the flawed, temporary, and corruptible body. Mendelssohn again: "As long as we trudge here on the earth with our body; as long as our soul is encumbered with this earthly scourge, we cannot possibly flatter ourselves to see this wish [for wisdom] completely fulfilled." The soul is almost always defined by its contrast with the body. Of course, all conceptions of personal immortality, rebirth, and reincarnation clearly require a soul that can exist outside of the body.

In most theological conceptions, the soul does not occupy a definite region of space—you can't put a physical box around it and say what's in the box is the soul and what's outside the box is not. (However, in Chinese philosophy, one version of the soul inhabits the liver *temporarily,* and in Descartes's philosophy, the soul inhabits the pineal gland in the brain, *temporarily.*) Another commonly invoked feature of the immaterial soul is that it cannot be subdivided. It's always a whole thing, as in Mendelssohn's argument.

The soul seems to be some kind of energy. But even if it's envisioned as energy, the soul would not be material in the modern scientist's understanding of energy. The various forms of energy discussed in physics—such as energy of motion, gravitational energy, electromagnetic energy—are generated by material particles. According to physics, all forms of energy occupy space, and the amount of energy in any given region of space can be measured and quantified. Furthermore, that energy can be converted into a very specific and quantifiable amount of matter, via Einstein's famous formula $E = mc^2$. Thus, the physicist's energy is part of the material world. Not so the soul.

In Chinese philosophy, each living being has two souls: a *hun,* which leaves the body after death, and a *po,* which remains in the body after death. The *hun* soul takes the form of three gentlemen, by the names of You Jing, Tai Guang, and Shuang Li, who live in the liver. At death, the *hun* becomes a *shen,* a word meaning "god" or "spirit," and which is associated with Heaven. We can also think of the *hun* in terms of the fundamental concept of Chinese thought: the yin-yang duality. The *hun,* essence/spirit/soul, is yang, in contrast to the body and Earth, which is yin. The Chinese debate among themselves whether the soul is immortal, but almost certainly it is not material.

Chinese seal script for *hun*

The notion of two different kinds of souls also finds expression in Hinduism. There is, first, a universal soul, which is pure, immutable, invisible, and infinite. When this universal soul enters a particular body, it becomes an individual soul, called the *atman* (or "self"). A section of the ancient Hindu text the Srimad-Bhagavatam puts it this way: "The spirit soul [universal soul], the living entity, has no death, for he is eternal and inexhaustible . . . [but] he is obliged to accept subtle and gross bodies created by the material energy and thus be subjected to so-called material happiness and distress."

The belief in two varieties of souls seems to nourish two different desires: the desire for personal immortality and the desire for an eternal, ethereal world of which we are a part. Most of us want our individual and personal selves to last far beyond a meager century or so. Existence and life seem too magnificent an experience to end when our material bodies disintegrate. At the same time, we find comfort in believing that some timeless intelligence or realm gives purpose to this strange universe we find ourselves in and embraces us individual beings.

Buddhists do not believe in a personal soul—a thing that would retain the identity of the individual being—but they do believe in an immortal and nonmaterial consciousness, which might be likened to the universal soul of other religious traditions. The current (and fourteenth) Dalai Lama calls this immortal consciousness "inner space." In a recent film titled *Infinite Potential,* the Dalai Lama describes this inner space as the deepest and most subtle level of consciousness, a kind of cosmic consciousness that is much larger than any individual living being. When a child is born, he or she inherits a piece of this cosmic consciousness. It has no beginning and no end. This cosmic consciousness, or inner space, is the only permanent thing in a universe that is otherwise impermanent. In fact, this cosmic consciousness precedes our particular universe. Universes come and go, come and go in endless cycles, but the cosmic consciousness persists.

To believe in a soul or a cosmic consciousness, whether the *ka* or the *shen* or the *atman* or the Buddhist inner space, we need to enlarge our idea of the universe. We have to imagine that beyond the world of atoms and molecules, tables and chairs, there's an ethereal world, which may include spirits, ghosts, Heaven and Hell, and an afterlife. In the next chapter, I'll talk about vitalism, the concept that living things have some nonmaterial essence, absent in nonliving things, that does not obey the laws of physics, chemistry, and biology. That vital spirit would also be part of the ethereal world. The ethereal world is nonmaterial. According to my definition of materialism—that the world is made of material stuff and only material stuff—belief in any one of the components of the ethereal world would constitute nonmaterialism. Worldwide today, the majority of people believe in various parts of this ethereal world. For example, according to the Pew Research Center, 72 percent of Americans believe in Heaven, defined as a place where people who have led good lives are eternally rewarded in some kind of disembodied existence. Fifty-eight percent believe in Hell. According to a survey of almost thirteen hundred adults by YouGov, 45 percent of Americans believe in ghosts. Essentially all of the 1.2 billion Hindus in the world believe in an immortal soul and some kind of reincarnation. Essentially all of the 1.8 billion Muslims in the world believe in an afterlife.

I never believed in any of these things. I'm not sure exactly why. I do know that from an early age, I developed a scientific view of the world—not especially from books but from my own experiments. In a large closet attached to my bedroom, I created a laboratory, which I stocked with beautiful glassware and chemicals and coils of wire. And I built things. I measured things. I constructed pendulums by attaching a fishing weight to the end of a string. I'd read in some book that the time for a pendulum to make a complete swing was proportional to the square root of the string. (For example, a pendulum of twenty-four inches should have a swing time twice as long as a pendulum of six inches.) I built lots of pendulums of different lengths and, with a stopwatch and a ruler, personally verified this amazing law. And it worked every time. As far as I could tell, nature behaved according to numbers and rules.

At age twelve, after seeing the Frankenstein movie with the fantastic giant electrical sparks flying about, I decided to build an induction coil of my own. This involved wrapping about a mile of thin wire around some metal rods. A thicker wire wound around the rods served to make them magnetic. When the electricity through that wire (produced by a simple six-volt battery) was rapidly turned on and off, the magnetic field of the rods oscillated. That, in turn, created a large electrical current in the thin wire, which then could produce leaping sparks. The oscillating magnetic field was, of course, invisible, but it produced highly visible effects. So, from my induction coil I learned that even invisible energy can be measured and obeys certain rules. I didn't see any reason to believe in some kind of mystical substance that couldn't be measured and managed.

An important experience that contributed to my commitment to the material world came one summer when I was visiting my grandparents at their seaside cottage. At night, I amused myself by walking out to the end of their dock and throwing stones into the water. One night, I had the urge to stir the ocean with a stick. To my surprise, it shimmered. I stirred again. Again the water glowed. I'd never seen such a thing. This seemed to be pure magic. I scooped up some of the "supernatural" ocean water in a jar and took it back to the house, to show Nana and Grandpoppy this magical thing I'd discovered. In the house, I peered more closely at the water in the jar and could see tiny bugs swimming around. So that was the source of the light! It wasn't magic. It was just little bugs. It was material stuff. I later learned that certain animals and plants have particular molecules that light up when agitated. It's called bioluminescence. Instead of being disappointed, however, I was even more delighted. Little bugs were capable of wondrous things.

Although as a child I developed a scientific view of the world, I also understood that not all things were subject to quantitative analysis. I remember some particular experiences I would now call spiritual, although

I wouldn't have used that vocabulary at the time. One remarkable experience stands out. I was about nine years old. It was a Sunday afternoon. I was alone in a bedroom of my home in Memphis, Tennessee, gazing out the window at the empty street, listening to the faint sound of a train passing a great distance away. Suddenly I felt that I was looking at myself from outside my body. For a brief few moments, I had the sensation of seeing my entire life, and indeed the life of the entire planet, as a brief flicker in a great chasm of time, with an infinite span of time before my existence and an infinite span of time afterward. My fleeting sensation included infinite space. Without body or mind, I was somehow floating in the gargantuan stretch of space, far beyond the solar system and even the galaxy, space that stretched on and on and on. I felt myself to be a tiny speck, insignificant. A speck in a huge universe that cared nothing about me or any living beings and their little dots of existence—a universe that simply was. And I felt that everything I had experienced in my young life, the joy and the sadness, and everything that I would later experience meant absolutely nothing in the grand scheme of things. It was a realization both liberating and terrifying at once. Then, the moment was over, and I was back in my body.

What was that thing I experienced at nine years old? Despite the dismal feeling that the universe didn't care a whit about me, I did feel connected to something far larger than myself. Perhaps such experiences, which many of us have, motivate the belief in a universal soul, or a universal consciousness.

Although Mendelssohn probably knew of the Egyptian and Chinese and Indian conceptions of the soul, he would have been more influenced by Western philosophy, his direct antecedent. Plato was a primary inspiration. In a passage from the *Phaedo* (ca. 360 BC), Socrates explains to one of his followers, using his usual method of a rhetorical question, that "the seen is the changing and the unseen is the unchanging?" (As usual in Plato's dialogues, Socrates slyly gets across his views by asking questions he knows the answers to.) Socrates goes on to say that the soul is the unchanging, unseen part of a living being, "dragged by the body into the region of the changeable . . . But when returning into herself she reflects, then she passes into the other world, the region of purity, and eternity, and immortality, and unchangeableness."

Plato clearly identifies that invisible, unchanging thing with the soul. It's remarkable to me that Plato could conceive of the invisible, certainly in the literal sense. Aristotle's five elements from which the cosmos was made—earth, air, water, fire, and aether—were all *visible*. Even air was visible. Air was the breath of a person on a cold winter day, or the mist rising from a pond in early morning. All visible. In ancient times, how did one imagine the invisible? How *could* one imagine the invisible? (Certainly, we can imagine things we don't currently see, like a book in a closed drawer, but a thing that is *unseeable* is different.) Since the nineteenth century, we've known that there are many invisible things. For example, only a narrow part of the electromagnetic spectrum is visible to the human eye. Gamma rays, X-rays, ultraviolet, infrared, radio waves are all invisible. The magnetic fields of my childhood induction coil were invisible.

And why does Socrates state, without any refutation by his disciple, that the invisible is unchanging? On what grounds? The invisible things that we know today—the X-rays and radio waves—do change. They move from place to place. They can be created and destroyed. Of course, we have the hindsight of two thousand years of scientific discovery. Still, Socrates is making some assumptions that could have been questioned by his followers (perhaps using the Socratic method in reverse).

In addition to its invisibility, the *indivisibility* of the soul is one of its defining attributes, closely related to its lack of extension in space. By contrast, anything material can be localized in space as well as divided into parts. Eight centuries after Plato, Saint Augustine (354–430), in his *Letters,* wrote that a soul could not be a thing that "occupies a larger place with a larger part of itself and a smaller place with a smaller part." By this statement, Augustine is not only saying that the soul does not occupy physical space, but also that one cannot think of the soul as having parts. A thing with parts can be divided, but the soul, as conceived by these thinkers, was an indivisible whole. Further, Augustine agreed with Plato that the soul is a thing in its own right, independent of the body: "The soul . . . seems to me to be a special substance, endowed with reason, adapted to rule the body."

Indivisibility is one of those qualities we associate with fundamental and perfect things. Perfection includes a notion of completeness, like a symphony that would be ruined should any note be deleted. Perhaps ironically, we human beings have always yearned for perfection even though we've never seen it. As I will discuss later, the ancient Greeks imagined that the world was made of tiny, indivisible things called atoms. However, the Greek atoms were material. Unlike souls, atoms occupied physical space. Today, we know that atoms can be split into even smaller pieces, protons and neutrons, which themselves are composed of even smaller things called quarks. In fact, one of the goals of modern physics, using our giant atom smashers, is to find the absolutely most fundamental and elemental particles of nature, which can no longer be split. The end of the line, so to speak. I suggest that this search for the indivisible and the perfect must be part of our human psychology—not only to better understand and predict the world that we live in but also to satisfy some deep longing for a world beyond, a world of perfection.

Skip another eight centuries after Augustine and we come to Saint Thomas Aquinas (1225–1274), arguably the most influential of all Christian thinkers. Saint

Thomas played a role in the rediscovery of Aristotle as his works were translated from Greek to Latin and tried to reconcile Aristotle's philosophy with Christian doctrine. But the reconciliation could be pushed only so far. Aristotle argued that the universe had no beginning, in sharp contrast to the account of Creation in Genesis.

In his discussion of the soul, Aquinas begins, as did most other philosophers and theologians, by associating the soul with life: "The soul is defined as the first principle of life in those things which in our judgment live; for we call living things 'animate' and those things which have no life 'inanimate.'" Aquinas goes on to argue that the soul cannot be material, because if the principle of life were intrinsic to bodies, then all bodies would be living things. Since some bodies, like rocks, are clearly not living, the starting assumption must be wrong. Thus, the soul is not intrinsic to material bodies; it must be something without body, that is, something immaterial.

If I understand this argument correctly, I find it to be logically in error. Why couldn't the "first principle of life" reside in some material bodies but not in others? Some leaves are rounded, but that does not mean that a pointy thing cannot be a leaf. Some leaves happen to be pointy. I am trying to refute Saint Thomas's rational argument for the immaterial soul with another rational argument, but, again, I believe that all rational arguments for the soul are on shaky ground. You either believe or you don't.

As with most Christian thinkers, Saint Thomas believed that the soul was immortal: "Some powers belong to the soul alone as their subject; as the intellect and the will. These powers must remain in the soul after the destruction of the body." To a modern scientist, immortality may be the hardest concept to accept. Nothing we know in the universe is immortal. Even stars eventually use up their nuclear fuel and turn into cold cinders floating in space.

I often wonder what it is that gives me a sense of myself, an ego, a self-awareness. Where does that sense come from? How does that unique sensation arise from mere atoms and molecules? How do thinking and emotion arise from mere atoms and molecules? Regardless of the answer to these questions, that we have thoughts is an undeniable fact. And from that fact, the philosopher René Descartes (1596–1650) constructed the world. *Cogito, ergo sum*. I think, therefore I am. In my view, this is the most powerful and convincing statement ever articulated by a philosopher. What is more controversial is Descartes's claim that the thing that has thoughts, the mind, is of a completely different nature from the material body—the so-called mind-body duality. The mind, for Descartes, is nonmaterial. In his *Discourse on the Method of Rightly Conducting the Reason and Seeking for Truth in the Sciences* (1637), Descartes says not only that the mind is immaterial but that it can exist independently of the body. Descartes's argument, in essence, is that he can conceive of a world in which he has no body but cannot conceive of a world in which he has no thoughts. "From that I knew that I was a substance whose essence or nature of which is to think, and that for its existence there is no need of any place, nor does it depend on any material thing; so that this 'me,' that is to say the soul by which I am what I am, is entirely distinct from body." By "soul," Descartes means the immaterial and unique essence of the particular human being, one of whose functions is to think. What distinguishes Descartes from the other philosophers and theologians we have considered is that Descartes starts with the mind, rather than the soul, although the two are related for him and both are immaterial. Before postulating an immaterial substance, he begins with something he knows for sure—that he is a thinking thing.

My objection to his argument for the separation between immaterial mind/soul and material body is similar to my problem with the argument of Saint Thomas. Simply because Descartes can imagine a bodiless world doesn't mean that his thinking mind inhabits that world. And he has not shown that his thoughts do not "depend on any material thing." The philosopher Rebecca Goldstein has pointed out to me a more subtle level of Descartes's argument. He claimed to understand that thinking was part of his essence, even without any experience of the physical world, and that this knowledge came to him long before he understood that he had a physical body. The problem with this claim is that some things have essential qualities (essences) that cannot be known before physical experience. For example, an essential quality of water is that it is made of atoms of hydrogen and oxygen. But we could not imagine that fact without experience with the physical world. We might imagine a thing we call "water," but if that imagined thing was not made of atoms of hydrogen and oxygen, it would not be water.

As I will discuss in the third chapter, modern neuroscience has good evidence that the thinking self is rooted in the material brain and nervous system—a physical basis for thought that Descartes could not have known, and certainly an essential quality of thinking that requires experience with the physical world. In other words, in the view of modern science, there is only a single substance, the neurons and atoms of the nervous system, although those neurons are capable of producing spectacular phenomena such as consciousness, self-awareness, imagination, and intelligence. I do believe in the materiality of the mind, but I confess that I am still mystified by the nature of consciousness.

Like others before him, Descartes says that a distinguishing characteristic of the soul is that it cannot be subdivided: "One cannot in any way conceive of a half or a third of a soul, or of what extension it occupies, and from the fact that [the soul] does not become smaller from some part of the body being cut off." Descartes differed from most of his predecessors' view that the soul gives life to the body. In his philosophy, the body was a mechanical thing that wore out (died) when its heat and motion were spent, whereas the immaterial soul was closely associated with thinking, a completely separate (nonmaterial) thing from the body.

Descartes's dualism is not supported by our modern understanding of biology. We now believe that all of our thinking occurs within the physical nervous system, even though we still do not yet understand the physical basis for consciousness. (See chapter 3.) Thus, in biology today, mind and brain are the same thing. However, as late as the 1950s, the prominent biologist John Eccles argued for a Cartesian-type separation between mind and brain. In a famous paper published in 1951 titled "Hypotheses Relating to the Brain-Mind Problem," Eccles claimed that experience, memory, and thought are "unassimilable into the matter-energy system." I find it fascinating that even well into the twentieth century, a leading biologist continued to believe in a nonmaterial basis for consciousness and thought, underscoring the totally unique sensation of consciousness and the mystery of the first-person experience.

Finally, as I try to grasp the antecedents of Mendelssohn, I come to Gottfried Wilhelm Leibniz (1646–1716), not only a leading philosopher but a highly distinguished scientist and mathematician. In fact, Leibniz published his invention of calculus in advance of Newton, although the latter actually made his discovery sooner. Mendelssohn, according to one of his biographers, regarded Leibniz as the greatest of all philosophers. The author of *Phädon* admired Leibniz's optimistic notion that our world is "the best of all possible worlds."

He also admired Leibniz's roots in science and mathematics, providing an analytical framework for Leibniz's thinking much like that of Mendelssohn himself.

Leibniz had his own view of the nonmaterial, something he called "monads." In his philosophy, monads were the indivisible elements that made up the world. They were infinite in number, although each was unique and acted independently of the others. They were shapeless. They were without length or width or breadth. Thus, they didn't occupy space. They were simple. They were not made of any material themselves, yet all material things were composed of them. Leibniz called his monads "the true atoms of nature." But they were very different from the atoms of the ancient Romans and Greeks, which were material and extended in space.

Every philosopher, from Confucius to Aristotle to al-Kindi, has attempted to understand the world in terms of fundamental elements. For Leibniz, those simplest, indivisible elements were the monads. Furthermore, it was through the monads that God created "the best of all possible worlds," because each monad was made by God and programmed with its own individual instructions to bring about harmony and perfection. However, the monads were not souls. As the simplest possible units, they did not have sensations and memories—which, according to Leibniz, were required for souls.

Today, little remains of Leibniz's monads. They're not atoms, because they're not material. They're not the abstract elements of mathematics, because each one is unique. And Leibniz himself said that they're not souls.

Still, the soul of the ancients is alive and well today. Most of the people I know, in all walks of life, believe in some nonmaterial thing that survives their bodily death. At an address to his General Audience in 2014, Pope Francis said that "Heaven, more than a place, is a state of the soul." Micah Greenstein, a prominent rabbi at Temple Israel in Memphis, Tennessee, told me that "we are not bodies with souls, but souls with bodies. The afterlife is the ultimate reconnection with God." When I talk to my wife about the soul, she says that she likes to keep her options open. And she feels pretty strongly in a nonmaterial cosmic energy that connects all living things. Such a cosmic energy seems related to the *ka* of the ancient Egyptians and the inner space of modern Buddhists. For those who believe in such cosmic energy, it must produce feelings like what I experienced when I made eye contact with those ospreys in Maine.

. . .

I've reread Mendelssohn's *Phädon*. And I'm struck by the number of times he mentions truth, wisdom, and perfection. I'm beginning to think that his striving for these qualities—even more than his desire for immortality or for a reunion with God—lay behind his passionate belief in the soul. Here are a few examples: "We are certain, that the knowledge of the truth is our only wish . . . We see clearly, that we will never reach wisdom, the goal of our wishes, until after our death . . . Do you see my friends! how far the man who loves wisdom must distance himself from the senses and their objects, if he wants to grasp . . . the All Maximum and Most Perfect Being?" Even at a young age, Mendelssohn aspired to knowledge and perfection and the life of a philosopher. When twenty-six years old, in an essay titled *Über die Empfindungen* (On Sentiments), he wrote that "the contemplation of the structure of the world thus remains an inexhaustible source of pleasure for the philosopher. It sweetens his lonely hours, it fills his soul with the sublimest sentiments . . . There lies in me an irresistible drive towards completeness and perfection."

Although few of us are philosophers, many of us embrace the ideals of purity, perfection, and wisdom. For Mendelssohn, the immaterial soul was not only the vessel for these ideals, but also the vehicle to achieve them—after we pass through this material realm. Mendelssohn was deeply devoted to God. He was a master of languages. He was a father and a husband. Above all, he wanted to learn how the world works, the truth of the world. Just like a scientist.

There are a number of reasons, I think, why so many of us believe in the soul and the ethereal world where it lives. Of course, there's the desire for continuing existence beyond our personal deaths. And, as for Mendelssohn, the yearning for perfection and purity. For many of us, as Rabbi Greenstein said, the desire is for reconnection with God. I'd suggest also the attraction of a place far removed from the dust and difficulties of the waking world we know. The longing for such a world is not unrelated to the belief in miracles, shared by a majority of people around the world. Why do we believe in miracles? We want to experience awe, wonder, novelty. The miraculous can be frightening, but it can also be exhilarating. Part of that appeal was stated nearly three centuries ago by the Scottish philosopher David Hume in his essay "Of Miracles" (1748): "The passion of surprise and wonder, arising from miracles, being an agreeable emotion, gives a sensible tendency towards the belief of those events, from which it is derived." In their book *Wonders and the Order of Nature,* the historians of science Lorraine Daston and Katharine Park document humankind's enchantment with wonders and oddities. Surprises and peculiarities. Miracles. Marco Polo enthuses over finding completely black lions in the Indian kingdom of Quilon. Other travelers excitedly record gourds with little lamblike animals inside;

beasts with the faces of humans and the tails of scorpions; unicorns; and people who vomit worms. The world of the miraculous, like the ethereal world, is a place of the imagination, a place not constrained by experienced reality.

I would guess that at times all of us want some escape from this humdrum and arduous life. Mendelssohn was born a hunchback. He was heckled and harassed for being a Jew. The world of the soul offered escape. There, he could disappear into the loving embrace of truth and perfection.

For me, mathematics has been such a world. It's a world of purity and perfection. It's a world of truth. It's a world of certainty, as clean and crisp as a new twenty-dollar bill. The circumference of a circle divided by its radius is *always* the same number, a particular number, with an *infinite* number of digits. When I visit the world of mathematics, sitting at my desk scribbling equations or reading a math book, I lose all track of my body. I lose track of time and space. Numbers and differential equations, curves, planes, and tetrahedrons are mansions in the clouds, solid and phantasmal at once. You can gaze out into that world, bodiless of course, and see all kinds of strange and wonderful things, and you have the feeling it's been there forever. Martians would understand. I can stay in that world for hours, until I get tired or need to eat. It's perfect. Maybe the soul is like that. Sometimes, I wish I believed in the soul. But I've got mathematics.

The soul, although not material, lives in some endless and infinite domain of time and of space, perhaps beyond time and space. Infinities may not exist in the physical world. We'll never know. What we do know is that nothing in our physical world endures forever. Everything eventually disintegrates and passes away. Cities crumble. Forests burn. Human beings deteriorate and die, their atoms disassembling and blending with soil and oceans and air.

There is, however, a cosmological theory, proposed by the distinguished Stanford physicist Andrei Linde, which predicts that our universe, and other universes, are constantly spawning new universes in an unending chain of cosmic creation, extending into the future for eternity. The theory is called "eternal chaotic inflation." It's backed up by some serious equations and has already made accurate predictions for phenomena in our particular universe. In some of his papers, Linde illustrates his eternal chaotic inflation model as a thick hedge of branching bulbs, each bulb a separate universe, connecting to ancestor bulbs and descendant bulbs by thin tubes. The entire collection of universes is called the multiverse. It's startling to look at Linde's picture and realize that each bulb represents an entire universe, some containing stars and planets, cities, trees, ants or antlike creatures, sunsets. Some probably pure energy, devoid of life. We'll probably never be able to determine whether Linde's theory is true. But it offers the possibility that our universe may not be all that there is, in space or in time.

And, thinking completely as scientists now, if we take a truly cosmic view of the world, beyond the life of individual people, even beyond the life of our particular universe, we might conceive of some kind of immortality, one of the attractive features of the immaterial soul. Individual universes, like individual lives, may come and go. But the constellation of all universes, branching off from one another, may go on forever.

2

Primordia Rerum

A Brief History of Materialism

In ancient Greece and Rome, death was as familiar as the neighbor next door. If a mother gave birth to ten children, only three would likely live to age ten. For childhood survivors, life expectancy did not extend beyond the midforties. A floor mosaic from Pompeii depicts the rooftop of a house, beneath which are emblems of domestic life: a wheel, an iron for poking the hearth fire, a sack of grain. In the middle is a large skull, the memento mori, the constant reminder of death.

With no understanding of germs, people had little idea of what waited to kill them. "Roman fever" was the name for malaria. Dysentery, another common killer, literally meant "bad intestines." Many young people died of mystifying infectious diseases such as typhoid, diphtheria, and influenza. Other major causes of death included venereal disease, cholera, plague.

Thucydides vividly described the plague (probably

smallpox or typhus) that struck Athens in 430 BC and killed a third of its population: "People in good health were all of a sudden attacked by violent heats in the head, and redness and inflammation in the eyes . . . When it fixed in the stomach, it upset it; and discharges of bile of every kind named by physicians ensued, accompanied by very great distress. In most cases also an ineffectual retching followed, producing violent spasms."

Medicines and attempted cures were based more on superstition than on science. For many ailments, bloodletting was the standard solution. Some physicians made their diagnoses by tasting the patient's bodily fluids. Epilepsy was treated by eating a dried camel brain soaked in vinegar. Although there was some awareness of the hazards of large gatherings in crowded spaces, the warnings were often ignored, further spreading infectious diseases. Apartment dwellers dumped their excrement into the street. Toilets in the market were simply holes in the ground. To clean themselves after using the public toilet, customers shared an implement called a xylospongium, which was essentially a sponge on a stick.

But there was something even worse than the fear of capricious suffering and death. That was the fear of what might come after. The ancient Romans and Greeks believed that the souls of the doers of bad deeds would be tortured in Hades forever. The darkest and most awful region of the underworld was called Tartarus. After death, a person's soul was brought before Rhadamanthus, a demigod and once a king of Crete. According to one of Plato's dialogues, if Rhadamanthus judged a person to have sinned, his soul "is marked with the whip, and is full of the prints and the scars of perjuries and crimes with which each action has stained him, and he is all crooked with falsehood and imposture, and has no straightness, because he has lived without truth." Those wretched souls found guilty of the worst crimes would endure "the most terrible and painful and fearful sufferings as the penalty of their sins"—for all of eternity. Virgil described Tartarus as "the sea of night profound, and punished fiends." He goes on to say that Tartarus is surrounded by triple walls, to keep its condemned residents from escaping, and one can hear "the groans of ghosts, the pains of sounding lashes and dragging chains."

A marble carving from southern Turkey, dating to the second century AD, depicts the eternal punishment of Ixion, former king of the Lapiths of Thessaly, who

murdered his father by pushing him onto some burning coals. The setting is Tartarus. Ixion is shown chained to a wheel, while another character flogs him.

Challenging the grim possibility of eternal torture and pain was a Roman philosopher and poet named Titus Lucretius Carus (ca. 99 BC–54 BC). The afterlife is pure superstition, said Lucretius. It doesn't exist. The body and soul are nothing but material atoms, which he called *primordia rerum,* "the first beginnings of things." When a person dies, her atoms scatter like "mist and smoke disperse abroad into the air . . . Therefore, death is nothing to us."

Lucretius was the most influential materialist of the ancient world. And his conception of atoms as the minute and indestructible building blocks of nature echoed through the centuries, later heard by Dalton and Einstein.

Lucretius borrowed his atomic hypothesis from the Greek thinkers Democritus (460 BC–370 BC) and Epicurus (341 BC–270 BC). Like them, his principal mission was to lessen the fear of death. Yet his expression of those ideas, a book-length poem of some seventy-four hundred lines called *De rerum natura* (On the Nature of Things), not only went beyond his Greek predecessors in its depth and insights, but also was regarded as a masterpiece of literary beauty and passion. The Roman orator Cicero wrote that Lucretius's poem was "full of inspired brilliance, but also of great artistry."

Early Christian churchmen disparaged the poem

because of its rejection of an eternal soul and religion in general. Then, the verses nearly disappeared for a thousand years. The poem would have been lost forever, except for the efforts of a fifteenth-century Italian scholar named Poggio Bracciolini (1380–1459), who discovered what may have been the last surviving copy of *De rerum natura* in a German monastery and brought it back into the world. Today, the poem is considered one of the greatest literary works in the Latin language.

There's much scholarship on *De rerum natura*. I'll comment on the portions that have meant the most to me. I first encountered the poem in college. Previously, in high school, I discovered that I had a miserable ear for spoken languages, so I took refuge in Latin. I never had to speak a word. On paper, I could proficiently conjugate verbs for hours at a time—*sum, es, est, sumus, estis, sunt*. Seeing my strengths and deficiencies, my teachers politely encouraged me to fulfill my college language requirement with Latin rather than German or French. As I remember, Lucretius was not assigned, as was Virgil and Catullus, but I read him anyway, to the bafflement and amusement of my roommates.

I was a physics major. Although I could quickly see that the author of *De rerum natura* was not up on relativity, I was mesmerized by the *completeness* of his scientific explanation of the world. Some atoms were smooth, some jagged; some were soft, some hard; some closely packed, some loosely assembled with the spaces filled by the "void"—and these atomic features explained all the different properties and behaviors of matter. Thunder was caused by the collisions of clouds. Lightning "when clouds by their collision have struck out many seeds of fire; as if stone or steel should strike a stone, for then also a light leaps forth." Every natural phenomenon in the cosmos found explanation in terms of atoms.

Because atoms could not be created or destroyed, the universe had always existed, from infinite time in the past. Also, the universe was infinite in space, without outer boundary. Lucretius reaches this conclusion by supposing that a brave volunteer goes to the extreme edge of the cosmos and throws a spear outward. Will it hit something solid or keep going? Either option presents a problem for a finite and bounded universe, said Lucretius. If the spear continues, then there is space beyond; if it is obstructed, then there must be material to do the obstructing, and that material occupies volume. So, there can be no end to space. Good reasoning. (Two thousand years later, Einstein's theory of the cosmos, and the way that gravity can alter the geometry of space, would invalidate Lucretius's argument.)

More than anything else, I was impressed by Lucretius's belief in a lawful nature. Not necessarily in quantitative terms, for there is not a single number or specific law in the poem, but in the notion that nature obeys logic and laws, outside the province of the gods, and those laws can be understood by mortal human beings. I became a fan.

Little is known of Lucretius. Evidently he followed the Epicurean advice to live quietly and unnoticed. One of the very few records of his life are some sentences in the *Chronicles* of Saint Jerome (342–420), who wrote, "The poet Titus Lucretius was born. He was driven mad by a love potion and, having composed in the intervals of his insanity several books which Cicero afterwards corrected, committed suicide in his forty-fourth year." Lucretius was probably an aristocrat, as suggested by his poem's familiarity with Roman life, including that of the privileged class. He addressed his poem to Gaius Memmius, a judicial officer in 58 BC whom Lucretius considered an equal.

De rerum natura consists of six books. The first two lay out the atomic hypothesis. Following Democritus, the world is conjectured to be made of tiny, indivisible, indestructible particles—the atoms, or "first beginnings." There are an infinite number of atoms of various sizes, shapes, and weights (all too small to be seen), giving rise to the different properties of different materials in the world.

Change occurs as a result of atoms rearranging themselves, *without the need for outside agency.* The atoms are eternal, while the objects made out of them are not. The third and fourth books concern the soul/spirit, which is "delicate and composed of minute particles and elements much smaller than the flowing liquid of water or cloud or smoke . . . since mist and smoke disperse abroad into the air, believe that that spirit also is spread abroad and passes away far more quickly, and is more speedily dissolved into first bodies [atoms]." The important point: the soul, like the body, is material.

Lucretius used the same word for mind, spirit, and soul: *animus* or *anima*. He argued for the materiality of the mind/spirit/soul because the mind develops with the body. For small children, says Lucretius, the mind is weak. It strengthens as the child grows older, then declines again in old age, when the body is "wrecked with the mighty strength of time." The link of the mind to bodily aging is an unusual and compelling argument for the materiality of the soul/mind.

The fifth book of *De rerum natura* explains the origin of the world, the movements of heavenly bodies, and the creation of human society—all in terms of material atoms. The last book concerns various natural phenomena such as thunder and lightning, again explained in material terms, and ends oddly and abruptly with a recounting of the plague that struck Greece in the fifth century BC.

While Lucretius's principal motive in his materialist conception of the world was to eliminate the fear of death, another intention was to free humankind from the vagaries of the gods. Since everything was made of atoms, and atoms could not be created or destroyed, the gods were severely limited in their power. They could not make things suddenly appear or disappear.

If I had been born two thousand years ago, fear of death and of the capricious acts of the gods would certainly have been uppermost on my mind. Some of those concerns are still prevalent today. A recent Pew survey indicates that more than half of Americans still believe in Hell, as a place where sinners are punished after death. The percentage may be larger worldwide. On the other hand, worries about divine intervention in human affairs have lessened with the rise of monotheistic religions, such as Christianity, Judaism, and Islam.

I regard Democritus and Lucretius as critical thinkers. Although they didn't perform experiments, they regarded the physical world as a domain of rules and self-consistent validity, much as I did with my pendulums as a child. They did not swallow the beliefs and broadcasts of their society without questioning them. Through their atomic hypotheses they were interested in explaining the *causes* of things, not just their attributes: everything that happens in the world has a cause, and that cause originates from the movement and properties of physical atoms, not the gods.

In his emphasis on causes and a mechanical explanation of the world, Lucretius differed markedly from Plato and Aristotle, who were more concerned with *purpose* than cause. Aristotle did indeed list four necessary causes (*aitia*) for things to happen: the initial material, a form for the material, an agent to bring about change, and an end or purpose. But all of these steps were guided by the final purpose, and the conception as a whole was far more abstract than Lucretius's atoms. Although some of Lucretius's science was wrong, his reasoning, like that of Moses Mendelssohn nearly two thousand years later, was thoroughly modern in its sensibility. Lucretius should be considered not only a poet and philosopher but also a scientist. He wanted to understand how the world worked in terms of basic principles.

One of the first materialists in the Asian world was the Chinese meteorologist and astronomer Wang Ch'ung (27–97). In his book *Lun Heng* (Discourses Weighed in the Balance), Wang takes a highly rational view of the world. He says that thunder is heat, not messages from the gods. He says that the belief in ghosts is erroneous. And he denies the existence of an afterlife, writing, "The souls of the dead are dissolved, and cannot hear any more what men say."

I am struck by the similarity between Wang's language ("the souls of the dead are dissolved") and the words of Lucretius ("spirit also is spread abroad and passes away far more quickly, and is more speedily dissolved into first bodies"). Of course, the original words in Chinese and Latin were not identical, but the concept was the same. Both Lucretius and Wang proclaimed that a living body has some kind of spirit, but that spirit is a material thing, and it dissolves and disperses upon death. Whereas with Plato and Saint Augustine and others, the spirit/soul was an immortal and ethereal substance that maintained some kind of identity beyond death.

The belief in (immaterial) ghosts was widespread in ancient China, but Wang gave a logical and witty counter-argument:

> From the time when heaven and earth were set in order and the reign of the "Human Emperors" downward, people died at their allotted time. Of those who expired in their middle age or quite young, millions and millions might be counted . . . If we suppose that after death a man becomes a ghost, there would be a ghost on every road and at every step . . . They would fill the halls, throng the courts, and block the streets and alleys.

Wang was born into a poor family in the northeast part of Zhejiang Province. Lacking the money to buy books, he is said to have spent much time reading in bookshops. It seems that from an early age, he rebelled against authority. After quarrels with various people, he resigned his position as an officer of merit and went into self-imposed exile, during which time he composed the various essays on philosophy, government, and morality that constitute the *Lun Heng*. At the center of Wang's thinking and work was a rejection of Daoism and Confucianism and a rejection of the state philosophies and their endorsements of divine power. In particular, Wang disagreed with the idea that the gods controlled human beings and had a purpose for us. In this regard, his thinking was similar to that of Lucretius. Both philosophers argued that a materialist and scientific understanding of the world would free us from the power of the gods. Perhaps a questioning of authority is a prerequisite for critical thinking, certainly in ancient times and probably today as well.

Another chapter in the early history of materialism concerns the explanation of vision. The eye is the most critical portal for receiving information from the outside world. How do we see? In ancient Greece, and for centuries afterward, two competing theories of vision held sway: the visual ray theory of the Pythagoreans, and the *eidola* theory of the atomists (Democritus, Epicurus, Lucretius). The Pythagorean school, which included such people as Plato and Euclid, argued that the light we see by is a divine fire, closely connected to the immaterial soul, originating in the body. The eye is like a flashlight. It sees the outside world by emitting light rays, which travel to objects, illuminate them, and reflect back into the eye. By contrast, the atomists believed that the light used in vision originates outside the eye. In this theory, all objects are continuously throwing off delicate but material images of themselves, called *eidola,* which travel to the eye and allow sight.

Definitively refuting the Pythagorean theory of vision was the Egyptian physicist Ibn al-Haytham (965–1040), a pioneer in the study of light. According to al-Haytham, light is an essential form (*sūra*) intrinsic

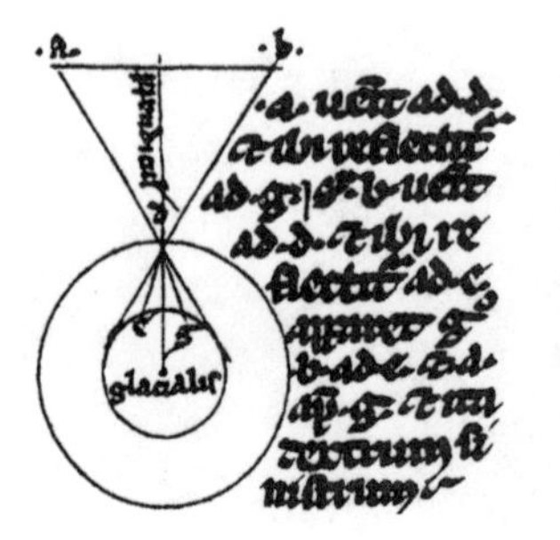

to self-luminous bodies. With his own experiments and mathematics, the Arab physicist worked out a theory of the trajectory of light rays, establishing a one-to-one correspondence between points on an external object and points inside the "crystalline humor" of the eye (the material filling the eyeball, behind the pupil).

Al-Haytham declared the Pythagorean hypothesis that vision is accomplished by light originating in the eye "superfluous" and "useless." As he sensibly observed, it would be absurd to think that light emerging from a person's eyes could illuminate the entire sky as soon as he opened his eyes.

In subtle ways, al-Haytham advanced the materialist view of the world. The body, including the eyes, might have various internal spirits, but there was a material world beyond the body, which created the light by which we see. And al-Haytham's optical drawings demonstrate the mechanics of that external, material world. Thus, he reinforced the notion that a physical world exists beyond the body and mind, a statement not unanimously agreed upon by philosophers until the twentieth century. Of course, the only things we directly know about the world are through our sensory perception and thoughts. That much is almost a tautology. We hear sounds, we see images, we feel surfaces, we process those sensations in our nervous system, and we infer an outer world. But to further claim that the world is totally a mental fabrication—as proposed by Bishop Berkeley and other philosophers—does not seem at all tenable to me. If that view were true, then we would never be surprised by what we find in the outer world. Yet we are constantly surprised. Galileo was surprised to see craters on the Moon when he first looked at that silver disk with his homemade telescope. Everyone thought the Moon would be perfect and unblemished, as befitted a heavenly body. Alexander Fleming was surprised when he found dead bacteria in his petri dishes after leaving them exposed to open air on his laboratory bench—the discovery of antibiotics. We are not sleepwalking through our lives. There is a material world beyond our bodies, made of light beams and mountains and other living things.

A long-standing debate in biology, bearing directly on the materialist conception of the world, has been over the difference between living things and nonliving things. It's called vitalism versus mechanism. The vitalist school of thought decrees that the transformation of nonliving matter to living matter requires some nonmaterial essence or vital force—outside the laws of chemistry, biology, and physics. Plato and Aristotle were vitalists. As was Descartes.

Until the mid-twentieth century, some prominent scientists shared the vitalist view as well. The influential French physician Paul-Joseph Barthez (1734–1806) argued that there were three different kinds of substances: matter (*la matière*), life (*la vie*), and soul (*l'âme*). Barthez claimed that the laws that govern matter could not necessarily be applied to the nonmaterial soul. Much the same statement was made by the distinguished Swedish chemist Jöns Jacob Berzelius (1779–1848). In the last edition of his *Lärbok i kemien,* considered the most authoritative chemical text of the first half of the nineteenth century, he wrote, "In living nature the elements seem to obey entirely different laws than they do in the dead."

About the same time as Berzelius published his textbook, Jean Antoine Chaptal (1756–1832), one of the important chemical industrialists of his age, wrote that "chemistry, in its application to living bodies, may therefore be considered as a science which furnishes new means of observation . . . But let us beware of intermeddling in the peculiar province of vitality. Chemical affinity is there blended with the vital laws which defy the power of art." This short passage reveals much. Most obviously, Chaptal agrees with Barthez and Berzelius and other vitalists that the laws that govern living things cannot be understood by "art," which here means science. More interestingly, Chaptal says we should not be "intermeddling" in the "province of vitality." The phrase "province of vitality" strongly suggests the existence of some ethereal world beyond the reach of science, beyond materiality, a world where we should not assume that every action has an equal and opposite reaction (as in the Newtonian world). That we should not "meddle" in such a world suggests that there are areas of knowledge forbidden to mortal man. I am reminded of the passage in Milton's *Paradise Lost* where Adam asks the angel Raphael how the heavens work. To explain the daily apparent rotation of the stars, is it the heavens revolving about a sedentary Earth, or is it the Earth that moves? Raphael replies:

> *This to attain, whether Heaven move or Earth*
> *Imports not, if thou reckon right; the rest*
> *From Man or Angel the great Architect*
> *Did wisely to conceal, and not divulge*
> *His secrets, to be scanned by them who ought*
> *Rather admire.*

That we should "beware" of such meddling, or admire rather than inquire, harks back to the sin of Adam and Eve in partaking of the tree of knowledge. Chaptal's comment implies that the mysterious realm of vitality, the realm of life force, is not only one we don't understand, but also one that we should *not attempt* to understand. The argument goes that there are boundaries to human investigations and knowledge, beyond which we are not entitled to explore. Such notions remain in the world today. Twenty-five years ago, when the first animal was cloned, a sheep named Dolly, some commentators argued that such a development was a sacrilege, human beings playing God.

In opposition to vitalism is mechanism, which holds that a living body is just so many biological pulleys and springs and chemical flows, with no metaphysical spiritus needed. For example, biologist Georges-Louis Leclerc, count de Buffon (1707–1788), rejected Newton's belief that God could and must intervene with the workings of the physical world. In his *Théorie de la terre, preuves,* Buffon wrote that "in physics one must, to the best of one's ability, refrain from turning to causes outside of Nature." And in his *Oeuvres philosophiques* he wrote that "life and animation, instead of being a metaphysical point in being, is a physical property of matter."

Through the history of biology, the vitalism-mechanism debate has weaved back and forth. The debate was brought into sharp focus in the nineteenth century, led by scientists in Germany. In particular, as the modern law of the conservation of energy was being articulated in the 1840s (more on that shortly), the chemists Justus von Liebig and Julius Mayer independently proposed that the energy needs of animals are supplied solely by the chemical breakdown of food, with no hidden energy provided by some internal and ethereal vital force. According to the mechanists, a gallop, a grinding of teeth, a warm breath on a cold winter's night could not possibly occur without ingestion of food. No more than a ball on level ground could begin rolling without a push.

In the late nineteenth century, German physiologist Max Rubner (1854–1932) began testing Mayer's and Liebig's hypothesis in more quantitative detail. Rubner made use of other recent work by chemists and nutritionists who had measured the chemical energy stored in various foods. Each gram of fat, carbohydrate, and protein has its energy equivalent. Rubner tabulated the energies required for body heat, muscle contractions, and other physical activities and compared the total against the chemical energy in food. By the end of the century, he concluded that the energy used by a living creature equals the energy consumed in food. In other words, the physicists' law of conservation of energy is also true for biology. There are no hidden sources of energy, no production of energy from nothing. On the ledger sheets of energy, a living being can indeed be considered a container of so many coiled springs, balls in motion, weights on cantilevers, and electrical repulsions.

I would add that the vitalist school in biology, a piece of the nonmaterialist view of the world, draws a sharp line between what is knowable by human beings and what is not. Historically, the unknowable lies within the exclusive province of God, or perhaps of enlightened beings. By contrast, the mechanist view is associated with the belief that there is only a single kind of substance in the cosmos, and that substance comprises both living and nonliving things. To be sure, living things have a special arrangement of their atoms and molecules that leads to the activity we call life. A bird is different from a stone. But in the view of modern biology, it is all material. Furthermore, according to the materialist view, the stuff of both life and nonlife can be understood by human beings, including all the laws that stuff obeys. Certainly, we don't completely know all the laws of nature now. But we believe that those laws are within our power to understand at some time in the future. All that said, as I will suggest in the next chapters, our purely material bodies are capable of extraordinary experiences, such as consciousness and spirituality.

In my opinion, the most decisive understanding of the world as material, and only material, has been accomplished in physics. A key part of this understanding is the notion that the world is a lawful place, following rules and cause-and-effect relationships. That notion of lawfulness is part of what attracted me to Lucretius a half century ago. One of the earliest quantitative laws of nature was formulated by the Greek scientist and mathematician Archimedes (ca. 287 BC–ca. 212 BC). As with Lucretius, little is known of Archimedes's life. But we do have a record of his law of floating bodies (ca. 250 BC): any solid object less dense than a fluid will, when placed in the fluid, sink to the level at which the weight of the displaced fluid equals the weight of the object.

We can speculate on how Archimedes arrived at

his law. At the time, balance scales were available for weighing goods in the market. The scientist could have first weighed an object, then placed it in a rectangular container of water and measured the rise in height of the water. The area of the container multiplied by the height of the rise would give the volume of water displaced. Finally, that volume of water could be placed in another container and weighed. Undoubtedly, Archimedes performed this exercise many times with different objects before devising the law. He probably also performed the experiment with other liquids, like mercury, to discover the generality of the law, which applies to all liquids.

Beyond its particularity to bodies floating in fluids, Archimedes's law suggested that the natural world followed predictable behavior—again the same conclusion I reached as a boy with my pendulums. Centuries later, this notion was reaffirmed by Italian physicist Galileo Galilei (1564–1642), considered one of the first modern scientists, in his law of falling bodies: the distance a body falls in the gravity of the Earth, or in any constant acceleration, is proportional to the square of the time to fall that distance. For example, if a body falls ten feet in one second, it will fall ninety feet in three seconds. The mathematical laws for floating bodies and falling objects were part of a growing conception of a rational, logical, lawful natural world. In such a world, there would be no place for ghosts and souls and other ethereal essences.

In 1609, at the age of forty-five, Galileo heard about a

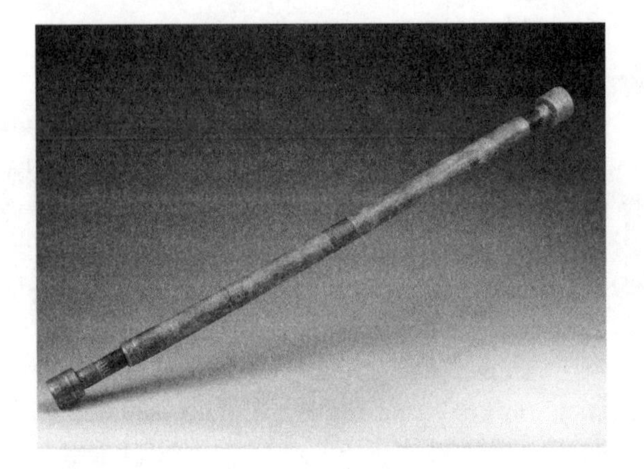

new magnifying device just invented in the Netherlands. Without ever seeing that marvel, he quickly designed and built a telescope himself, several times more powerful than the Dutch model. It seems that he may have been the first human being to point such a thing at the night sky. (The telescopes in Holland were called "spyglasses," leading one to speculate on their uses.)

I recently saw one of Galileo's original telescopes in the Museo Galileo in Florence, Italy. It is a remarkably simple device, consisting of a reddish-brown tube about forty inches long, an eyepiece at one end, and a lens at the other. It is made of wood, paper, and copper wire. When I peered out of an exact replica of the telescope, I was surprised at how small the field of view was, about half the width of the Moon. And I could barely make out what I was looking at. The images were dark and

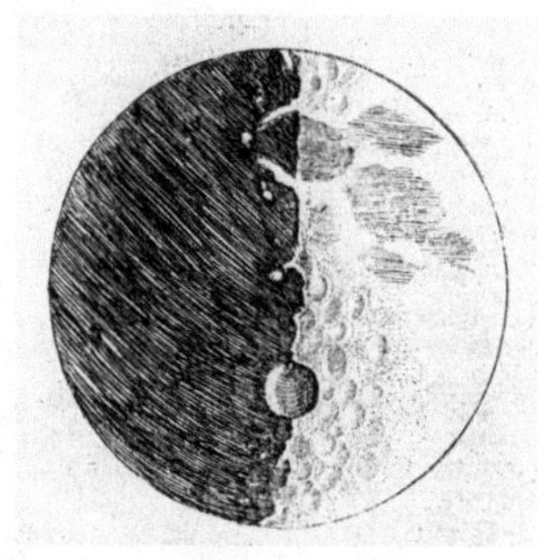

dim, due to the small amount of light coming through the lens and down the tube. Evidently, Galileo had to let his eyes adapt to very low light.

What Galileo saw with his telescope were jagged features on the surface of the Moon and transient spots on the Sun (now known to be relatively cool areas produced by the Sun's chaotic magnetic field).

In his little book *Sidereus nuncius* (Starry Messenger), Galileo exhibits his own pen-and-ink drawings of the Moon seen through his telescope, showing dark and light areas, valleys and hills, craters, ridges, mountains. He even estimates the height of the lunar mountains by the length of their shadows.

When he peered at the dividing line between light and dark on the Moon, the so-called terminator, it was not a smooth curve as would be expected on the perfect sphere of theological belief, but a jagged and irregular line. "Anyone will then understand," Galileo writes, "with the certainty of the senses that the Moon is by no means endowed with a smooth and polished surface, but is rough and uneven and, just as the face of the Earth itself, crowded everywhere with vast prominences, deep chasms, and convolutions." And in a letter on May 12, 1612, to the Italian scientist Federico Cesi, Galileo wrote of his observation of dark patches on the Sun, called sunspots. Galileo went on to convey his disdain for the armchair philosophers and theologians: "I wait to hear spoutings of great things from the Peripatetics [school founded by Aristotle] to maintain the immutability of the skies."

These observations and comments by Galileo challenged the prevailing notion that the heavens and heavenly bodies were composed of some indestructible and divine substance, as with the immaterial soul. That divine substance, sometimes called the "ether" or "aither" or "primary body" or "fifth element," was described by Aristotle as "eternal, suffering neither growth nor diminution, but ageless, unalterable and impassive." The word "ethereal" is, in fact, derived from the Greek word for ether, *aitheras*. It was only a short leap from Galileo's observations to the conclusion that the heavens, moons, planets, and stars are made out of the same material as Earth. The heavens are not heavenly. We live in a material cosmos.

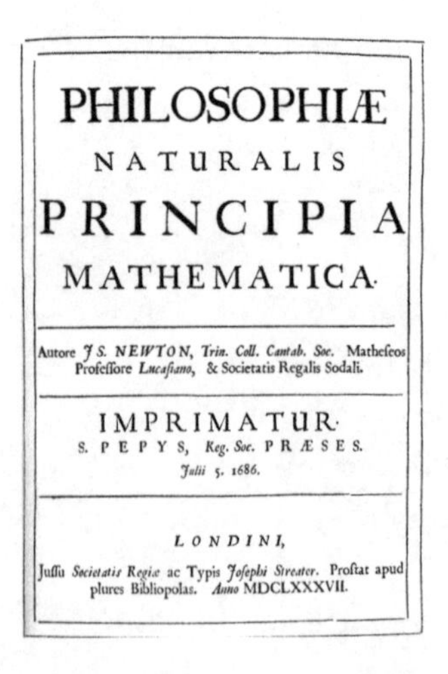

PHILOSOPHIÆ
NATURALIS
PRINCIPIA
MATHEMATICA.

Autore *J S. NEWTON*, *Trin. Coll. Cantab. Soc.* Mathefeos Profeffore *Lucafiano*, & Societatis Regalis Sodali.

IMPRIMATUR.
S. PEPYS, *Reg. Soc.* PRÆSES.
Julii 5. 1686.

LONDINI,

Juffu *Societatis Regiæ* ac Typis *Jofephi Streater*. Proftat apud plures Bibliopolas. *Anno* MDCLXXXVII.

The work of Isaac Newton (1643–1727) must surely be considered a landmark in the emerging concept of a lawful universe. Newton's law for gravity was not only one of the first mathematical expressions of a fundamental force underlying the motions of bodies. It was also the first proposal that a rule for the behavior of material bodies on Earth should apply in the heavens as well—that is, the first real understanding of the universality of a law of nature. Part of Newton's brilliance was the recognition that the same force that caused an apple to fall from a tree also caused the Moon to orbit the Earth.

His law of gravity can be stated this way: the strength of the gravitational force between two masses doubles when either mass doubles and increases fourfold when the distance between them is halved. Newton's law explained in quantitative detail the motions of the planets—the elliptical shapes of their orbits, the speeds and how they varied, and much more. Newton's laws of motion and law for gravity represented a giant step forward in our understanding of the cosmos and the natural laws that govern its behavior. The Nobel Prize–winning twentieth-century physicist Richard Feynman marveled that nature could obey "such an elegant and simple law as this law of gravitation."

Newton elucidated the nature of the gravitational force. In 1865, the Scottish physicist James Clerk Maxwell (1831–1879) did the same for the electromagnetic force. Maxwell and others before him showed how electricity could create magnetism and vice versa, so that the two were really part of the same phenomenon, like the famous drawing that appears to be either an old woman or a young woman, depending on how it's looked at. The Scottish scientist published a con-

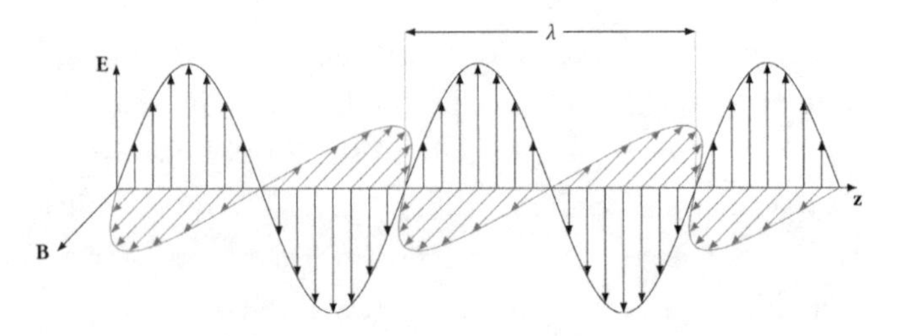

nected group of four equations, now known to all physics students as "Maxwell's equations," that completely spelled out the behavior of the electromagnetic energy field.

One particular prediction of Maxwell's equations was the existence of oscillating waves of electrical and magnetic energy that travel through space at the speed of light. (Indeed, visible light is such a wave.) Beginning in 1886, the German physicist Heinrich Hertz (1857–1894) began building equipment that generated and detected radio waves, which are electromagnetic waves of lower frequency than visible light waves. That work confirmed the existence of Maxwell's electromagnetic waves.

Energy is the stuff that drives the world. The importance of the discovery of electromagnetic waves, and the quantitative understanding of them through Maxwell's equations, cannot be overemphasized. The immaterial soul, the vital force of living things, and indeed the entire ethereal world, have often been associated with some kind of nonphysical energy, not analyzable by science. In the decades before Maxwell, there was a growing understanding by physicists of the concept of energy and how different forms of energy could be converted into one another: a book falling off a shelf moves faster and faster as it approaches the floor, evidently converting some kind of energy associated with its height (gravitational energy) into the energy of speed (kinetic energy). A hot gas can push against a piston and raise a brick as the gas cools, evidently converting some kind of energy associated with heat into the increased gravitational energy of an elevated weight. What Maxwell and Hertz discovered was that energy could actually move through space, going from one place to another. Not only that, but it could be localized, and it could be quantified. Thus energy, the thing that drove the cosmos, was not a mysterious essence that appeared willy-nilly, without rule or reason. It obeyed laws. It bowed to cause and effect. *Energy is part of the material world.*

Crucial to the understanding of energy was the idea that it cannot be created or destroyed. Although one kind of energy can be converted into another, the total amount of energy in a closed box is constant. We call this idea "the conservation of energy." It is one of the sacred cows of science. That the total amount of energy in an isolated region is constant harkens back to the ideas of Democritus and Lucretius that atoms are indestructible. Atoms cannot be created or destroyed; thus the total number of atoms in a closed box is constant.

Historically, heat played a crucial role in formulating the idea of the conservation of energy. Gravitational energy and kinetic energy (energy of motion) are clearly related, since any dropped object picks up speed as it approaches the surface of the Earth. By contrast, the conversion between these forms of energy and heat energy is not so obvious. Heat consists of the random motions of microscopic particles and is not visible to the naked eye. The realization in the nineteenth century that heat was a form of energy, able to change into other forms of energy according to a strict accounting, gradually led scientists to the notion of the conservation of total energy.

The German physician Julius Robert Mayer (1814–1878) is credited for first proposing an equivalence of all forms of energy, including heat, and the conservation of total energy. One might naturally ask how a physician discovered the energy equivalent of heat.

It's a fascinating story. Mayer, born in Heilbronn, attended the classical gymnasium school there, transferred to the evangelical theology seminary at Schöntal, and then enrolled in the medical school at the University of Tübingen. There, he received his doctorate of medicine with distinction in 1838. In early 1840, the twenty-five-year-old Mayer boarded a Dutch merchant ship on a voyage to the East Indies, where he served as ship's physician. Then, in Java, while letting the blood of sailors (for who knows what medical treatment), Mayer was surprised and impressed by the redness of their blood. He correctly reasoned that the heat of the tropics allowed a lower rate of metabolic combustion to preserve body heat and thus required less oxygen from the red blood cells, accounting for their unusual redness. It was already known that the oxygen combined chemically with foodstuffs to produce energy. Mayer concluded that chemical energy is related to animal heat, and that the conversion between the two could be expressed quantitatively. He then went on to generalize the equivalence of all forms of energy. In his pioneering essay published in *Annalen der Chemie und Pharmacie* in 1842, Mayer writes that "Forces are causes . . . In numberless cases we see motion cease without having caused another motion or the lifting of a weight; but a force once in existence cannot be annihilated, it can only change its form . . . If, for example, we rub together two metal plates, we see motion disappear, and heat, on the other hand, makes its appearance . . . From this point of view we are very easily led to the equations between falling force, motion, and heat." Although Mayer uses the word "force," he is actually talking about what physicists now call "energy," which is force acting through a distance.

Once scientists had conceived of the concept of the conservation of energy, they could then do a calibration experiment in which they measured how much heat energy was produced by a certain weight falling a certain distance. That would establish a quantitative measure of heat energy and its equivalence to a certain amount of gravitational energy. The same could be done in a calibration experiment to see how fast a mass was moving after falling a certain distance, establishing a quantitative measure of kinetic energy and its equivalence to a certain amount of gravitational energy. But these were only calibrations. After such initial investigations, physicists *might* have found that additional experiments produced many varying results in terms of how much heat energy was generated by a given amount of kinetic or gravitational energy—violating any notion of a general law of conservation of energy. However, that did not happen. What physicists actually found was that after establishing the equivalence of different forms of energy in the calibration experiments, those equivalences *always* applied in future experiments. With the proper equivalencies between different forms of energy, the total energy in a closed system was constant. The conservation of energy is evidently a law of nature. Energy may change forms, but the total cannot increase or decrease.

In the twentieth century, we have discovered and quantified two more forces of nature, the so-called strong force, responsible for holding the atomic nucleus together, and the weak force, responsible for the disintegration of certain subatomic particles. In all of these more modern developments, the law of conservation of energy has held. It is part of the great body of knowledge of modern physics that has convinced me, and many people worldwide, that nature is lawful. In the millions of experiments and probes of the physical world, modern scientists have not found any mysterious forces or phenomena that did not ultimately yield to a rational explanation. Sometimes, we have had to revise our understandings and theories based on new observations and considerations, as happened in the twentieth century with relativity and quantum mechanics, but the revised theories have always been consistent with the law of the conservation of energy and, more generally, a material, rule-based understanding of nature.

Even as a college student, I appreciated Lucretius for more than his atomic theory of the world. He was concerned with people. For starters, his principal motive for supporting the atomic hypothesis, like that of Epicurus, was to relieve the dread of death in his fellow human beings. "If men saw that a limit has been set to tribulation, somehow they would have strength to defy the superstitions and threatenings of the priests." Today, with modern medicine and more sophisticated beliefs about Heaven and Hell, it is hard to imagine the psychological trauma of such dread in an age when death came early and unexpectedly and many people faced the possibility of unending torture and pain in an afterlife.

Like Epicurus, Lucretius was also concerned with deep philosophical issues, such as the question of whether the decisions and actions we take are of our own choosing or instead already determined by the inevitable movement of atoms, mindlessly obeying cause-and-effect relationships back into the distant past. If atom A bumps atom B, atom B will bump atom C, and so on. In other words, are we simply robots? To allow ourselves freedom of choice, Lucretius proposed that the atoms, as they are moving downward through the void, are "at times quite uncertain and [in] uncertain places, they swerve a little from their course . . . such as to break the decrees of fate, that cause may not follow cause from infinity, whence comes this free will in living creatures."

Lucretius also had the beautiful—and scientifically correct—idea that our atoms were once part of other people, living before us, even though we have no memory of those ancestral arrangements of atoms. For me, the notion that our atoms were once part of other people and will again become part of other people after we die provides a meaningful connectedness between us and the rest of humanity, future and past.

In his thinking, Lucretius considered not only his fellow men and women on Earth. He had a cosmic view of life. Lucretius was one of the first philosophers to postulate the probable existence of life elsewhere in the cosmos: "When abundant matter is ready, when space is to hand . . . then you are bound to confess that there are other worlds in other regions and different races of men and generations of wild beasts." The idea of living beings on other planets, called cosmic pluralism, was suppressed by Christian doctrine for hundreds of years but then renewed in medieval times by such Islamic thinkers as Muhammad al-Baqir (676–733) and later in the Christian world by Giordano Bruno (1548–1600).

. . .

A big difference between materialists and nonmaterialists is in their attitudes toward death. Our inescapable death may be the single most powerful fact of our brief existence in this strange cosmos where we find ourselves. Indeed, one could argue that much of our thinking, our view of the world, our artistic expression, and our religious beliefs involve coming to terms with this fundamental fact. Such an argument is made by the cultural anthropologist Ernest Becker in his landmark book *The Denial of Death* (1973), where he asserts that our entire civilization is a defense mechanism against the knowledge of our inevitable deaths.

I find it fascinating that both materialists and nonmaterialists have been motivated by this same elemental fact, yet propose very different psychological strategies for confronting it. The nonmaterialists, such as Socrates and Saint Augustine, argue that we should not fear death. Actually, we should welcome it—because our immortal and immaterial soul will enjoy a blessed existence forever (if we do good). In Plato's *Phaedo,* just before Socrates swallows the hemlock, he tells his followers that "I do not grieve as I might have done, for I have good hope that there is yet something remaining for the dead, and, as has been said of old, some far better thing for the good than for the evil." And in his book *On the Trinity,* Augustine writes, "As, therefore, all men [desire] to be blessed, if they [desire] truly, they [desire] also to be immortal; for otherwise they could not be blessed." Belief in some kind of afterlife, as a defense against death, does not seem to have been much diluted by the advances of science. As mentioned in the last chapter, even today, 72 percent of Americans believe in Heaven, a place "where people who have led good lives are eternally rewarded."

By contrast, as we have seen, the materialists Epicurus and Lucretius argue that we should not fear death—because upon dying we dissolve. After death, we do not exist in any form. When we are nothingness, there is nothing to fear.

Driven by the same profound fact of our inevitable death, the materialists and the nonmaterialists have developed totally different means of eliminating the fear of it, in accordance with their different worldviews. What lies behind those differences in worldviews? One simple hypothesis is that those people who embrace a largely scientific view of the world are materialists, while the others are nonmaterialists. Although some of Lucretius's claims about the natural world were wrong—for example, he argued that the Earth was flat rather than round—he was scientific in his thinking. He sought physical explanations for all the phenomena of the world. The basic feature of the science of Lucretius and Epicurus is that all things are made of physical atoms, and only atoms, and that feature precludes the afterlife (and any fear of the afterlife).

Yet, this hypothesis for distinguishing materialists from nonmaterialists cannot be entirely correct. As previously discussed, some prominent scientists up to the twentieth century, such as Paul-Joseph Barthez and Jean Antoine Chaptal, argued that living things possess some *nonmaterial* essence not found in inanimate things. And a relatively recent survey conducted by the University of Chicago revealed that 58 percent of the physicians in the United States believe in some kind of afterlife (requiring a nonmaterial existence). Although that percentage is somewhat less than for the public at large, it is still quite high.

So, the reasons underlying a materialist versus a nonmaterialist view of the world must be more complex than simply the presence or absence of scientific thinking. Rather than attempt a complete explanation of these divisions, I'll again explore some of the thoughts and desires behind nonmaterialism—thoughts and desires that all of us experience to varying degrees; evidently, they hold less sway for materialists. First is the deep desire for permanence—against all of the evidence presented to us by nature. Everything we see around us in the natural world ultimately passes away. In the summer months, mayflies drop by the billions within twenty-four hours of birth. Forests burn down, replenish themselves, then disappear again. Ancient stone temples and spires flake in the salty air, fracture and fragment. And just look at our own bodies. In the middle years and beyond, skin sags and cracks. Eyesight fades. Hearing diminishes. Bones shrink and turn brittle. More personally, in the last decade my height has shrunk by over an inch. (I was never of great stature to begin with.) The natural world fairly shouts out to us that all is impermanent.

Yet we long for something that lasts, that endures beyond the shifting sands of accident and mortality. We associate permanence with *meaning*. As Becker says, our art, our religions, our nation-states are all attempts to create something that lasts, and we attribute meaning to things that last. We also associate permanence with divinity and perfection. We can readily see that we ourselves are imperfect things made out of clay (or atoms). But we aspire to perfection. Perfection is a manufactured idea. Nothing we see around us is perfect. Indeed, it takes a human mind to conceive of perfection, just as it takes a human mind to label something beautiful. The concepts of God and other divine beings are part of the perfection we imagine and seek. If everything material is impermanent, this reasoning goes, there must be something nonmaterial for permanence and its associated qualities of perfection and divinity.

Second is the inability to imagine nothingness. The materialists tell us that there is a moment in time when we will no longer exist, for infinite time into the future. How can anyone imagine such a thing? We cannot imagine nonexistence before we were born, and we cannot imagine nonexistence after we die. It is hard for us to believe that this spectacular and unique sensation we call consciousness—our thoughts and sensations, our smelling a bit of cinnamon or touching the velvety surface of moss—will someday come to an end. The experience of being seems too grand to be limited to a short four score and ten years or to our flimsy compartments of sinews and bone mass and blood.

A third factor, I suggest, is the special nature of living things. Living things simply do not behave like nonliving things. Rocks do not grow and reproduce. Soap bubbles do not evolve in a way to make them better able to survive hot days or tempestuous winds. Living things clearly have a number of special properties that led such vitalists as Aristotle and Descartes and Berzelius to conclude that they possess some nonmaterial substance outside the confines of physics and chemistry and biology. That said, essentially all biologists today are mechanists. A few years ago, I visited the laboratory of Harvard biologist and Nobelist Jack Szostak, who is attempting to create a living cell from scratch—in particular, from the simple molecules present in the primitive Earth. Szostak told me, "I hope that when we succeed, it will eventually seep into the culture that the creation of life is totally natural, and we don't need to invoke anything magical or supernatural." My own guess is that whether or not Szostak and his colleagues succeed in creating life from scratch, many people worldwide will still believe in some kind of nonmaterial essence or vital spirit present in living things. For example, the belief in chi, the life force that reputedly runs through our bodies, and all the various practices based on chi are part of a vitalist view of the world.

Lastly, as discussed in the last chapter, we are mesmerized and delighted by magic and miracles—phenomena beyond the world of appearances. Spirits and vital essences and immortal souls live in that ethereal world. The hieroglyphics carved on the walls of the temple of Unis in ancient Egypt were magical incantations to help the dead pharaoh ascend to Heaven. Even today, according to the Pew Research Center, 79 percent of Americans believe in miracles—events that lie outside natural law and any explanation by science. Not just the parting of the Red Sea or the resurrection of Jesus or the splitting of the Moon by Muhammad, but supernatural phenomena in the world of today, such as ghosts, voices from the dead, instructions from God, accurate prophecies, sudden recoveries from grave illnesses, telekinesis, reincarnation and more. Ross Peterson, a psychiatrist practicing in the Boston area, told me, "We want miracles as a solution to helplessness. We want miracles for meaning at a deeper level. Miracles lift us out of a humdrum life."

For all of these reasons, belief in a nonmaterial, ethereal world is deeply appealing and resonates with many of our psychological needs and desires. The materialists among us probably must have some particular experiences—such as my childhood activities with pendulums and bioluminescence, or the strong influence of a parent or teacher, or a natural pragmatic skepticism, or a disappointment in the hoped-for gifts of the ethereal world—to arrive at our materialist view.

From time to time after college, I have reread *De rerum natura*. As I've learned more science, I have come to better appreciate the widespread meaning and application of the atomic hypothesis. Long ago, I misplaced my original copy of the book, but I replaced it with an edition published in 1982 by Harvard University Press. It's a small red book about the size of a hand. It sits on my bookshelf next to others in the Loeb Classical Library series, including Boethius, Catullus, Euripides, and Augustine. Near these books are lighter fare: the poems of Emily Dickinson, several of Edgar Rice Burroughs's sci-fi romance novels set on Mars, Michael Ondaatje's memoir *Running in the Family,* Virginia Woolf's *Mrs. Dalloway,* Feynman's *The Character of Physical Law,* Rilke's *Letters to a Young Poet*.

As much as the science, I appreciate more and more the deeply human dimension of *De rerum natura*—above and beyond the bedrock of material atoms. Although we know little of Lucretius's life, we do know what he valued, as shown in his poem. And that masterwork demonstrates that he embraced many of the ideas and feelings I associate with spirituality. He certainly valued the happiness of other people, which he tried to increase by his reasoned argument against the fear of death.

He valued friendship, as evidenced in the way that he addressed Memmius: "It is your merit, and the expected delight of your pleasant friendship, that persuades me to undergo any labour." He valued living a good and moral life: "So trivial are the traces of different natures that remain [in people] . . . that nothing hinders our living a life worthy of gods." He had a sensitivity to beauty, as shown in this passage: "Golden images of youths about the house . . . cross-beams paneled and gilded echo the lyre, when all the same stretched forth in groups upon the soft grass beside a rill of water under the branches of a tall tree men merrily refresh themselves at no great cost, especially when the weather smiles, and the season of the year besprinkles the green herbage with flowers." And this expression of awe: "As soon as the brightness of water is laid in the open air under a starry sky, at once the serene constellations of the firmament answer back twinkling in the water." Like me, Lucretius was a spiritual materialist.

3

Neurons and I

The Emergence of Consciousness from the Material Brain

His love of dogs may have been the beginning of Christof Koch's interest in consciousness. "I've wondered about dogs since early childhood," he said in an interview a few years ago. "I grew up in a devout Roman Catholic family, and I asked my father and then my priest, 'Why don't dogs go to heaven?' . . . They're like us in certain ways. They don't talk, but they obviously have strong emotions of love and fear, hate and excitement, of happiness."

Dr. Koch, now chief scientist at the Allen Institute for Brain Science in Seattle after nearly three decades as professor of cognitive science at Caltech, is one of the world's leaders in studying the material basis for consciousness. Koch believes that consciousness is abundant. "It's much more widespread than we think in nature," he says. In fact, Koch and his collaborators in information science believe that someday nonbiological machines will be conscious.

The large question before us is, How can transcendent experiences arise from a material brain? I will suggest that such experiences, and others I have grouped together under the rubric of "spirituality," naturally emerge from a high level of *consciousness* and intelligence. In this chapter I'll consider the manner in which that consciousness might arise from a material brain. Part of this exploration is a study of the association between awareness and physical neurons of the brain. Part is to connect behavioral manifestations of consciousness to material brain structures, a connection particularly evident when the brain is damaged. Part is to show a continuum of such manifestations of consciousness through the animal world, from one-celled organisms (clearly not conscious) to rodents to chimpanzees to humans. Part is an attempt to identify a group of qualities needed for consciousness and to ask what kinds of material systems (living or not) might possess those qualities. And finally, I will examine emergent phenomena—collective behavior of complex systems not present or understandable in their individual parts—and view consciousness in the brain as such a phenomenon.

I will here concentrate on the brain, but, as the neuroscientist Antonio Damasio and others have argued, consciousness probably involves the entire nervous system and its integration with the full body.

However we define consciousness, it is probably a

graded phenomenon. It ranges from automatic responses to the surrounding environment at the low end to self-awareness, ego, and the ability to plan ahead at the high end. Amoebas may not be conscious in any meaningful way, while crows and dolphins and dogs almost certainly are. I will try to distinguish between these different levels of consciousness, but I may sometimes use the word "consciousness" more generically.

Let me acknowledge straight off that the highest level of this unique phenomenon, the primal human experience we call consciousness—the first-person participation in the world; the awareness of self; the feeling of "I-ness"; the sense of being a separate entity in the world; the simultaneous reception and *witnessing* of visual images, sound, touch, memory, thought; the ability to conceive of the future and plan for that future—all of that is so unique, so hard to describe, so different from experiences with the world outside of our bodies that we may never be able to fully capture consciousness with brain research. We may never be able to show in a step-by-step manner how this highest level of consciousness emerges from the neurons and synapses of the material brain. That is not to say that such an emergence does not occur. We may be able to show that the feelings and attributes we call consciousness are generated by material structures in the brain without being able to fill in all the blanks.

As I have said, I am a materialist. Like almost all modern biologists, I believe that consciousness and all mental experiences are sensations brought about by the chemicals and electrical currents in the brain. But we may never be able to cross the first-person/third-person divide. The experience of consciousness, at least at its higher levels, is the first-person subjective par excellence. The analysis of three pounds of neurons sitting on the lab table, the probing of that brain with instruments, the measuring of its electrical quiverings, the writing of equations to describe it, even the mere talking about it as a *thing* are all third-person activities. We can't be inside the box and outside the box at the same time. Of course, in some sense we are always inside the box of our own minds, since we cannot experience the world except through our individual brains.

In his famous 1974 paper "What Is It Like to Be a Bat?" the American philosopher Thomas Nagel defines consciousness in such a way that underscores the near impossibility of crossing the first-person/third-person divide: "Fundamentally an organism has conscious mental states if and only if there is something that it is like to *be* that organism . . . We may call this the subjective character of experience." How can we feel what a bat feels or what a dog feels or even another human being? In the words of the Finnish neuroscientist and philosopher Antti Revonsuo, "Nothing we can think about or imagine could make an objective physical process turn into or 'secrete' subjective, qualitative 'feels' . . . The best we can reach is a theory stating that, yes, consciousness does emerge from the brain, and then to simply list all the correlations between these two realities: when brain activity of type Z occurs, then a conscious experience of type Q emerges, and so on."

Some scholars argue that it is not simply the first-person/third-person divide that forever prevents us from understanding the emergence of consciousness. It is a physical limitation of our cognitive capacity. In his book *The Mysterious Flame,* the British philosopher Colin McGinn postulates that there is some "hidden structure" of consciousness that we don't understand at all. McGinn says that consciousness is an experience fundamentally beyond the capability of the human mind to fathom: "What we have in effect discovered over the centuries is that certain problems lie within the scope of our cognitive faculties, while others do not." McGinn, who is almost certainly a materialist, says that "our intelligence is wrongly designed for understanding consciousness" and even goes so far as to say that a different architecture of our brains might allow such an understanding. On that point, I disagree. A different architecture could not, in principle, solve the problem of the first-person/third-person divide. Consciousness requires information storage, and information storage requires material stuff, whether computer chips or neurons or confined patterns of electromagnetic fields. One still has the conceptual difficulty of getting from that material stuff to the first-person experience of consciousness.

Even with these difficulties, I still side with modern

biology in the belief that consciousness is an outcome of the material brain. That is, the mind and the brain are one and the same. In contradiction to Descartes, most modern scientists believe that there is only one type of substance in the universe, and it is a material substance.

Christof Koch was born to German parents in 1956, in the American Midwest. As his father was a diplomat, young Koch grew up in Holland, Germany, Canada, and Morocco. He studied physics and philosophy at the University of Tübingen, in Germany, and received his PhD in biophysics at the Max Planck Institute in Tübingen. In 1986, after four years at the Artificial Intelligence Laboratory at MIT, Koch joined the new Computation and Neural Systems program at Caltech. He has authored more than three hundred research papers and five books in cognitive science. He holds five patents. In addition to being a leading researcher in neuroscience, Koch spends some of his time communicating science to the public. He writes a regular column on consciousness for *Scientific American Mind* and gives public lectures. His 2004 book *The Quest for Consciousness* is a model of excellent popular science writing.

In 2005, Koch and a student invented a medical technique called "continuous flash suppression." A static image is presented to one eye, while a series of images is flashed to the other. Even though the static image is going to the brain, it eventually loses visibility; that is, the subject is no longer aware of the static image.

This experiment shows that visual consciousness requires some high level of processing beyond the incoming ocular information. The visual signals themselves do not contribute to awareness of a seen image.

Such a finding is likely related to the fact that only some of the neurons in the brain are involved with consciousness. Many activities of neurons are associated only with unconscious behavior, like reacting to a hot stove or automatic breathing. Humans who have lost most of their cerebellum due to stroke or trauma show no signs of having lost any consciousness. That consciousness is associated with some neurons and not others is further evidence for its material basis.

Another name for awareness is "attention." In the millions of visual images, sounds, smells, and other sensory inputs that bombard the brain every second, what mechanism allows us to pay attention to some things and to disregard others? What happens in the brain that enables us to ignore a leaking faucet but pay attention to a knock on the door? In 1990, Koch and Francis Crick (co-discoverer of the structure of DNA), building on an earlier idea of German neuroscientist Christoph von der Malsburg, proposed that our paying attention to a sight or sound is associated with the synchronous firing of neurons. Attention is not consciousness. However, it is probably a necessary condition for consciousness, and its neural mechanics are a step along the way to understanding the material basis of consciousness.

The attention proposal was supported in 2014 by neuroscientists Robert Desimone and Daniel Baldauf. These researchers presented a series of two kinds of images—faces and houses—to their subjects in rapid succession, like passing frames of a movie, and asked them to concentrate on the faces but disregard the houses (or vice versa). The images were tagged by flashing them at two different frequencies—a new face image every two-thirds of a second and a new house image every half second. The researchers then put a helmet-like device on the subjects' heads that could detect tiny local magnetic fields inside the brain and thus localized brain activity, a technique called magnetoencephalography (MEG). Another technique, called functional magnetic resonance imaging (fMRI), measures brain activity by the different magnetic properties of blood with oxygen (high activity) and blood without oxygen. By monitoring the frequencies of the magnetic and electrical activity of the subjects' brains, Desimone and Baldauf could determine where in the brain the house and face images were being directed and processed. The scientists found that even though the two sets of images were presented to the eye almost on top of each other, they were processed by different places in the brain: the face images by a particular region on the surface of the temporal lobe, the fusiform face area, known to specialize in face recognition. And the house images by a neighboring region, the parahippocampal place area, known to specialize in place recognition.

Most importantly, Desimone and Baldauf found that the brain cells (neurons) in the two regions behaved differently. When the subjects were told to concentrate on the faces but to disregard the houses, the neurons in the face location fired in synchrony, like a group of people singing in unison, while the neurons in the house location fired like a group of people singing out of synch, each beginning at a random part of the song. And when the subjects concentrated on houses and disregarded the faces, the reverse happened. Evidently, what we perceive as paying attention to something originates, at the cellular level, in the synchronized firing of a group of neurons, whose rhythmic electrical activity rises above the background chatter of the vast neuronal crowd.

Related to attention and the synchronous firing of neurons is the idea that *coalitions* of neurons compete with one another for our attention. We are usually not aware of these competitions. However, when one of the coalitions dominates the others, we become conscious of its message. For example, when we are trying to remember someone's name, we may go for seconds or minutes or even hours struggling to remember it. Unconsciously, lots of different coalitions, with different suggested names and perhaps visual images, are competing. Then, later, the correct name suddenly comes to us. At that point, one of the coalitions has won out over the others.

Desimone, who is the director of the McGovern Institute for Brain Research at MIT, thinks that we may be unnecessarily mystifying the experience we call consciousness. "As we learn more about the detailed mechanisms in the brain," he says, "the question of 'What is consciousness?' will fade away into irrelevancy and abstraction." As Desimone sees it, consciousness is just a word for the mental experience of attending, which we are slowly dissecting in terms of the electrical and chemical activity of individual neurons. He throws out an analogy: Consider a careening automobile. We might ask, Where inside that thing is its *motion*? But we would no longer ask that question after we understood the engine of the car, the manner in which gasoline is ignited by spark plugs, and the movement of cylinders and gears.

I visited Dr. Koch by Zoom in July 2021. At the time, both of us were spending what we hoped were the waning months of the coronavirus pandemic on small islands, his in the Pacific Northwest and mine in the Casco Bay region of Maine. His outdoor office is an expansive wooden deck, its perimeter lined with plants in turquoise and grey terra-cotta pots—the deck itself high on a hill with the tree-covered land sloping down on all sides. On this particular day, he wore a purple jacket and an ascot around his neck. He has silky grey hair and glasses. He speaks precisely, with a rather thick

German accent. While clearly serious about his clinical work on the material brain, he conveys a friendly human warmth and appreciation for the wonder of existence. "Life itself is such a mystery," he told me. "Just now, I can look at these trees, they are so vivid. The blue sky. I can smell the flowers. It's extraordinary. We take it completely for granted, because it's always been there. Then when we get into this very special space, where you feel in communion with the rest of the universe, then you think that's special."

In addition to doing laboratory experiments, like his work on continuous flash suppression, Koch and his collaborators have developed theories of consciousness and its hypothesized requirements. As far as the difficult first-person/third-person conundrum presented by consciousness, he told me, "So how do I know that you have feeling? I can use what scientists use all the time: induction. Given everything I know about brains and genes and evolution, it would be extremely unlikely for your brain to be very similar to mine but not have consciousness. And then I do the same for babies and for people who are paralyzed. And I do it for dogs and for cats. Now when I come to radically different systems, like squids and octopi that are very different from me, that don't have a cortex and look very different from my brain, it becomes more difficult. When I come to single cells, or trees, it becomes almost impossible. Then I come to computers. Then I need a theory. I need a fundamental theory that tells me a priori, which systems feel like something and which do not."

Following modern biology and neuroscience, we now believe that the activity of the brain takes place in neurons and in the interactions between them. We mostly understand how neurons work. A neuron consists of three parts: a cell body, which contains, among other things, the DNA of the cell; the dendrites, which are fibrous extensions of the cell that receive electrical impulses from other neurons; and the axon, which is a long slender projection from the cell body that transmit electrical signals to other neurons. The manner in which electricity flows through the axons is understood as an exchange of electrically charged atoms through the axon membrane. Individual neurons put out spikes of electrical discharges, about one-tenth of a volt and lasting about one-thousandth of a second. The manner in which messages are communicated from one neuron to the next is understood as a flow of certain chemicals, called neurotransmitters, across a tiny region between neurons called a synapse. All of these things have been observed, measured, and quantified.

There are about 100 billion neurons in the human brain. Each neuron connects to about a thousand other neurons, although the number varies from one part of the brain to another. Thus, there are about 100 trillion synapses in the brain. Human beings have the largest number of neurons of any known animal, except for the

African elephant and some whales. Jellyfish have about 6,000 neurons; ants have about 250,000; mice about 71 million; ravens about 2 billion; gorillas about 33 billion; whales about 150 billion; and elephants about 260 billion.

Are whales and elephants "smarter" than we are? Probably not, even though they are pretty smart. The most important measure of intelligence is likely not the absolute number of neurons, but the brain weight per body weight. Larger bodies require larger brains simply to manage all the nerve endings and internal functions, irrespective of intelligence. Thus, a more accurate measure of interanimal intelligence is something neuroanatomists call the "encephalization quotient": a comparison of the brain weight of a particular species against a standard brain weight of animals belonging to the same taxonomic group and with the same average overall body weight. By this measure, human beings are the "smartest" animals in our taxonomic group, with a brain weight 7.5 times the average mammal with our body weight.

Complex brain activity and consciousness is associated not only with the total number of neurons but also with the number of connections between neurons. As evidence: in the human brain, the cortex has fewer neurons than the cerebellum but many more connections between those neurons. From observing the association between behavioral manifestations of consciousness and damage to the cortex, neuroscientists have concluded that consciousness is much more associated with the cortex than with the cerebellum. The latter is responsible for much unthinking activity such as swallowing, and its neurons mostly act independently of one another. By contrast, neurons in the cortex have a great deal of interaction and feedback between themselves. As stated earlier, a person can lose much or all of her cerebellum and still show all the signs of consciousness. Not so with the cortex.

When we look at cortical neurons and the connections between them, humans surpass whales and elephants. However, there is one animal with even more cortical neurons than humans: the long-finned pilot whale, actually a kind of dolphin. It has twice as many cortical neurons as humans, 34 billion compared to 16 billion. So why aren't these dolphins more advanced than *Homo sapiens*? Possibly because they do not have hands to manipulate their environments, to record histories and write instruction manuals, and so on. That exception aside, all of these findings help confirm a material basis for consciousness. Further, they support the notion that higher intelligence and consciousness arise out of the interdependence of large numbers of neurons, like the synchronous firings in Desimone's experiments.

In terms of the numbers of neurons and connections between them, the complexity of the brains of the more "intelligent" animals is truly staggering. The largest computer simulations of brains have about two million "digital neurons," far fewer than in a mouse. But not only far fewer. In these simulations, each neuron is represented as a point, with no structure and no internal state—basically either on or off. By contrast, real neurons have variable inputs and outputs in the form of variations in the electrical potential across the cell membrane, and each neuron connects to a thousand other neurons. We are a long, long way from being able to simulate a brain on a computer.

Although consciousness almost certainly depends on large numbers of neurons working in concert, individual neurons can be amazingly specific in their activity. The behavior of a single neuron can be measured by inserting into it a tiny glass tube filled with potassium chloride, a liquid that responds to the electrical output of the neuron. Some neurons, called grandmother neurons,

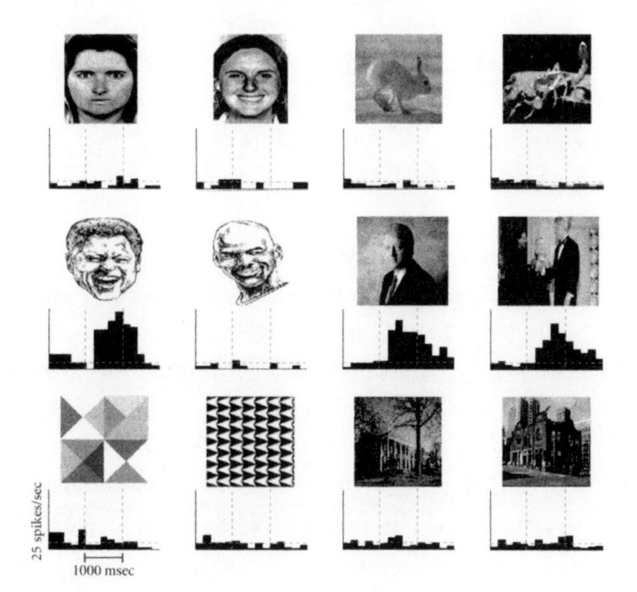

respond to the images of only particular people. (The images, of course, are seen by the eye and then transmitted to the brain.)

The figure above shows the electrical activity of one particular neuron (in a living patient's brain) that responds to pictures of Bill Clinton and nobody else. The electrical activity of the neuron is shown below the picture presented to the volunteer's eye. As can be seen in the top row, there is little response to pictures of a non-Clinton person, a rabbit, and a non-face. In the second row, the neuron responds vigorously to a cartoon of Clinton, a portrait of Clinton, and a group photo of Clinton, but little response to the non-Clinton face. The third row shows that there is little response of this particular neuron to abstract designs or buildings. Almost certainly, the information encoding Bill Clinton's face requires more than a single neuron, probably a group of neurons, but the individual units of that group are all highly selective and specific in their activity.

When Koch was a young researcher in his early thirties, he began a collaboration with the famous molecular biologist Francis Crick to understand consciousness. Together, they developed a list of minimal brain attributes needed to produce consciousness. Koch and Crick did not try to explain subjective experience, a sense of self, and the higher-level aspects of consciousness. Rather, they took a more modest approach of elucidating the neural requirements of just a single aspect of consciousness, visual awareness. For example, when we see a dog, what happens inside the brain for us to associate that visual input with the *concept* of "dog"? The minimal set of neurons involved in such a process Koch and Crick call "neuronal correlates of consciousness," or NCC. Such a list is the following:

- Neural complexity (greater than one hundred thousand neurons)
- Distributed nervous system, but highly integrated
- Lots of different types of neurons and different brain regions
- Sensory organs, including vision, with inputs to the brain
- Mental maps (one can navigate through space even when no sensory stimuli for guidance are present); neurons arranged to map outside world
- Neural hierarchies, with neuron-neuron interactions
- Many reciprocal, nonlinear connections between neurons
- Mechanism for selective attention
- Memory storage

The animals on planet Earth who have this full set of NCC are vertebrates, arthropods, and cephalopod mollusks (octopus, squid, cuttlefish).

Although Koch and Crick's goals were modest, we

can see that their NCC might apply to higher levels of consciousness beyond simple visual awareness. Self-awareness, both as a first-person "I" and as a being separate from the surrounding world, is one of the qualities we associate with a higher level of consciousness. Thus, a mental mapping of the outside world should be one of the neural correlates of consciousness—which would include awareness of one's position in time and space, both at the present moment and in the immediate future.

In 1970, the British American neuroscientist and psychologist John O'Keefe discovered particular cells in the hippocampus region of the brain, now called place cells, that fire when the animal is in a particular place. O'Keefe hypothesized that place cells might actually represent a physical map of the outside world. Then, in 2005, a Norwegian husband and wife team of neuroscientists, Edvard Moser and May-Britt Moser, discovered a group of cells in the entorhinal cortex, now called grid cells, that appear to integrate information about bodily location in space, distance, and direction. By attaching electrodes to rats' brains and seeing which neurons fire when the rats move about in open areas, the scientists discovered that the neurons fire only when the rat is at certain locations, and that those locations map out equilateral triangles in space. Thus, there seems to be an exquisite connection between the action of these cells and the animal's position in space. If the place cells are the brain's map, the grid cells are the brain's coordinate system.

Further evidence to support the relevance of a mental map of the world as an NCC comes when we consider human beings with compromised grid cells. It has long been known that older adults have difficulty with spatial navigation—a problem most acute with Alzheimer's patients who lose their way, even in familiar neighborhoods. Using fMRI techniques, neuroscientist Matthias Stangl at UCLA and other researchers have recently found that grid-cell activity in older adults is significantly reduced compared to younger adults.

All of these considerations further strengthen the understanding of consciousness as emerging from the constellation of neurons and connections between them. As Koch writes in *The Quest for Consciousness,* "Understanding the material basis of consciousness is unlikely to require any exotic new physics, but rather a much deeper appreciation of how highly interconnected networks of a large number of heterogeneous neurons work."

An important strategy in attempting to understand how consciousness emerges from the material brain is to identify outward manifestations of consciousness, what have been called "behavioral correlates of consciousness" (BCC), and to see whether these BCC are modified in people with brain damage, such as traumatic injuries (falls and car accidents), lesions and tumors in the brain, and strokes. One can further look for BCC in lower animals and attempt to trace some kind of evolutionary history of BCC.

Outward manifestations of consciousness include a sense of self, a distinct personality, memory, an ability to imagine the future and plan ahead, an awareness of one's mortality, a sense of play, an ability to solve problems, and others. Some of these attributes, such as problem-solving ability, might be associated with higher intelligence in general.

Psychiatrists, psychologists, and neuroscientists have developed questionnaires for patients with brain damage to measure their degree of self-awareness and functioning ability. The questionnaires are administered to three groups: the patient, the patient's family, and an observing clinician. One such questionnaire, developed by Mark Sherer at the Baylor College of Medicine and the University of Texas McGovern Medical School at Houston, asks such questions as:

- How well does the patient get along with people now as compared to before his/her injury?
- How well can the patient do on tests that measure thinking and memory skills now as compared to before his/her injury? . . .
- How good is the patient at keeping up with the time and date and where he/she is now as compared to before his/her injury?
- How well can the patient concentrate . . . ?
- How well can the patient express his/her thoughts to others now as compared to before his/her injury?

- How good is the patient's memory for recent events now as compared to before his/her injury?
- How good is the patient at planning things now as compared to before his/her injury?

Not surprisingly, the results of such studies show low scores *as measured by the family members and clinicians* but not particularly low by the patients themselves. Evidently, when a patient loses self-awareness, it is not so apparent to the patient themself. Such awareness of one's own lack of awareness would require another, supervisory part of consciousness unaffected by the brain injury. It is also possible that patients with brain injury are defensive about their loss of abilities and overrate their mental capabilities. Self-reporting is always a tricky business. Thus, most reliable here are the reports of family members and clinicians.

Autobiographical memory is an important feature of self-identity and self-awareness, for several reasons. Part of our sense of who we are is embodied in our store of memories—the events of our past lives. Part also stems from our social interactions at the present moment. Those interactions are much impacted by our autobiographical memory. Imagine going to a cocktail party with strangers under the restriction that you were not allowed to say anything about your past. The only thing the stranger can know about you is who you are at that moment: what you are wearing, your physical appearance, your knowledge of current events, your ability to carry on a conversation. Most of us would find such an experience difficult—not only uncomfortable but unsatisfying.

Numerous studies have shown that autobiographical memory is diminished by brain damage and dementia. Consider, for example, Alzheimer's, a disease that destroys memory and thinking ability. Autopsies of the brains of Alzheimer's patients reveal deposits of a protein called amyloid around brain cells, and plaques of another protein called tau that cause tangles of brain cells. Researchers have also found that as brain cells become affected in Alzheimer's disease, there is a decrease in the chemical neurotransmitters like acetylcholine that send signals between neurons. These findings not only show the clear correlation between memory (and associated consciousness) and the physical brain but also emphasize the importance of the communication between neurons as a critical part of consciousness and higher intelligence in general.

Accounts by people who are at the beginning stages of dementia, with still enough cognitive ability remaining to describe their situation, offer a harrowing glimpse into that experience. Here is such an account by Leo from Tasmania:

> When leaving my doctor's surgery one day I could not find my car. I did not know where I was. Eventually, I was able to make my way home . . .

> Independence has featured strongly throughout my life. I now depend on my wife, Ellie, to oversee any decisions I make. I find this very difficult. Timing and how you say things is very important, but my sense of timing is now lost. Say it now or forget it. People have disappeared from my life. It is like going through a divorce. I fear making a fool of myself.

Of course, it isn't necessary to study people with traumatic brain injuries or strokes to document the connection between the material brain and modified states of consciousness. We voluntarily alter our brains and resulting consciousness with various psychoactive drugs, such as alcohol, Prozac, Ritalin, marijuana, cocaine, MDMA, psilocybin, and LSD. It is known, for example, that LSD changes the way neurons communicate with one another by attaching to one of the proteins called serotonin receptors, which work with the neurotransmitter serotonin. A person with screen name El Charro Loco posted the following description of their experience after taking LSD:

> All of a sudden there was a disconnect between my thoughts and reality. I was no longer mindful of the present moment and it was like my consciousness was being deflected and drawn to something else, but I don't know what it is. I stopped my meal and came over to the computer to document

all this because I did not want to miss the take off in these notes . . .

I'm slowly starting to feel how I become more intertwined in a perception of reality that I hadn't experienced in a while, like if I had been keeping something locked in a basement and is now coming out. The feeling is not that of euphoria or wrath, there's nothing aggressive about it, it's more like going back to the house you were raised in and bringing out thoughts, behaviors and memories you hadn't had the chance to interact with in a while. Visual reality is starting to bend significantly. Hallucinations are starting to kick in at this point of the trip . . . Waves in the sound merge with colors in the air, the atmosphere of the entire room has a very smooth and pleasant rocking motion, like a baby in his mother's arms. My mind is not in the same place at this point . . . Slight shivers run thru my body as I type these words . . . I am not who I was a few hours before, at least not in terms of mental presence. Goliaths in my head battle for power from right to left in this constant dance of melodies along with the music. Rhythmic sounds and melodies resonate in the back of my head like the echo thru an empty hall.

El Charro Loco clearly has enough self-awareness left to be able to sit at the computer and record their experiences, using the pronoun "I," but their sense of time and space has been altered, as well as their memories.

One way of exploring the emergence of consciousness in the human brain is to study the behavioral correlates of consciousness in other animals and map out a gradation of consciousness with increasing brain capacities. It is almost certain that nonhuman animals have conscious experiences like humans. Few things in nature are all or nothing. There is always a continuum. Dolphins, which have almost as many cortical neurons as humans (the long-finned pilot whales actually have more), have shown clear signs of self-awareness and play. In a famous experiment demonstrating self-recognition, a mirror is placed in a pool with dolphins. The dolphins swim up to the mirror, look at it for a few moments, and swim away. Then marks are placed on the dolphins' bodies. Now, the dolphins spend longer looking at themselves in the mirror. Evidently, they have noticed that something has changed about their bodies.

Out in the open ocean, dolphins will stop what they are doing when a large boat approaches and ride in its bow wave. Some years ago, I went sailing in the Aegean Sea. A dolphin not only swam alongside us but catapulted itself over the stern of the boat. To all appearances, it was having fun. Monkeys play. Kittens chase one another and paw at a hanging string. Sea lions will toss sticks to one another. A hilarious video on YouTube shows a few minutes in the lives of some young crows. At first, the birds appear bored. Then one of them spots a low-hanging branch on a tree, flies up and grabs the branch, and swings back and forth on it. Nothing accomplished. But . . . The other crows notice how much fun their friend is having and come over and join in, taking turns swinging on the branch.

Problem-solving is certainly associated with intelligence and probably with higher levels of consciousness as well. The brains of corvids (crows, ravens, jays, magpies), although small, have neurons that are much more densely packed than in other animals. Consequently, these birds have as many neurons in their "bird brains" as some monkeys. And they manifest the behavior to prove it. A video posted in February 2021 shows a raven named Bran who is confronted with a box containing a meat treat. To get the treat, the bird has to perform a sequence of tasks in a certain order: (1) push aside a ball in front of the box, (2) withdraw three horizontal rods locking the entrance to the box, (3) drop a latch securing the door of the box, (4) open the door with a piece of string, and (5) reach inside the box and pull on another piece of string attached to the meat treat. Bran accomplishes all these tasks with great dispatch.

Experiments with chimpanzees have shown that they are able to gather information and base decisions on that information. Neuroscientist and psychologist Michael Beran and colleagues at Georgia State University did the following experiment with chimpanzees who had already been trained to choose pictures of familiar items on a keyboard. A particular food item was placed within an opaque container, sometimes with the chimps watching and sometimes not. The chimps were then rewarded with the food if they could name it on their keyboards. If the chimps had not seen the food item placed in the container, they would first go to the container and look before selecting that item on their keyboard. If they had already observed the food item being placed in the container, they would not inspect the container before choosing the item on their keyboards.

Awareness of mortality would seem to be a sign of higher-level consciousness and intelligence—not simply when a sick animal goes into a corner to die, but awareness of death within a social context. James R. Anderson and colleagues in the Department of Psychology at the University of Stirling in the UK took videos of the behavior of a group of chimpanzees during the late illness and death of an elderly female in the group named Pansy. After Pansy lies down and begins labored breathing, two of the other chimps begin stroking and grooming her. A third chimp then shakes her arm. One of the chimps strokes Pansy's hand. After Pansy gives a final twitch, indicating death, a male chimp jumps into the air and pounds Pansy's torso, then runs away. Pansy's daughter Rosie sits beside the dead body all night. The next day, the surviving chimps are profoundly subdued.

From all of these examples, various levels of consciousness certainly seem to be present in nonhuman animals. Higher levels of consciousness probably require the ability to manipulate the environment, to record history and information, and to pass that information on. For these actions, manual dexterity may be needed. An animal may be highly intelligent, but its perception of the outside world and of itself is much diminished if it cannot manipulate that world and record large amounts of information.

Psychiatrist and neurologist Todd Feinberg and biologist Jon Mallatt argue that the neuronal correlates of consciousness are confined to vertebrates, arthropods (insects), and cephalopods (squids, octopi, etc.). If so, it is possible to hypothesize when a primitive form of consciousness first emerged during the history of life on Earth. That would be about 540 to 500 million years ago, when the earliest fossils of these animals were found in rocks of the Cambrian period, during the so-called Cambrian explosion. This fairly rapid development in the evolution of life was probably caused by the appearance of the first predatory animals, odd-looking marine animals called anomalocaridids possessing good eyesight and a pair of grasping arms near their mouths. With the arrival of predators, other animals needed to adapt defensive mechanisms, requiring more advanced brain capabilities such as accurate (and quick) awareness of their position in space and the ability to plan and anticipate. Darwinian forces would have selected out those animals whose brains made such adaptations.

Just as there is a gradation of intelligence in animals, surely there must be a gradation of consciousness as well. Rats do not have the self-awareness of crows, and crows probably lack the self-awareness of humans. One might order levels of consciousness in the following way:

Rudimentary Life → First-Level Consciousness → Second-Level Consciousness → Human Consciousness

---

Rudimentary Life: Meets minimal requirements for life (e.g., microorganisms)

First-Level Consciousness: More complex systems, but always behaving in a reactive mode (e.g., worms)

Second-Level Consciousness: Higher intelligence, exhibiting signs of self-awareness, awareness of mortality, indulging in play, puzzle-solving ability, ability to predict (e.g., dogs, dolphins, chimpanzees, crows)

Human Consciousness: Still higher intelligence, creation of art, science, predictive powers, advanced manipulation of environment, etc.

The recognition of levels of consciousness in other animals, and an associated evolutionary history of brain capacities, further confirms the notion that consciousness is rooted in the material brain. And that the human brain and its capacities are not qualitatively different in kind from the brains of other animals.

In 2004, the Italian American neuroscientist Giulio Tononi, who holds the Distinguished Chair in Consciousness Science at the University of Wisconsin, pioneered a mathematical theory of consciousness called "integrated information theory" (IIT). Since then, Tononi has collaborated with Koch to further develop the theory. Unlike Koch's program of the neuronal correlates of consciousness, which begins with the brain, IIT begins with the *experience* of consciousness, categorizes the essential qualities of that experience, and then explores the kind of mathematical and material structures that would be needed to produce those qualities. According to this line of thinking, consciousness would not necessarily require biological neurons. It could emerge from any physical system, including a computer, that possessed the appropriate structure. As Koch says, "If you build a neuromorphic brain that has . . . let's say copper instead of axons, and transistors instead of neurons, then if you have the same cause-effect repertoires as [the] human brain, then this entity would actually be conscious." However, even according to IIT, consciousness requires a material structure that can act and produce change.

Tononi and Koch propose five qualities of conscious experience: (1) a thing exists from its own perspective;

(2) each experience is composed of many distinctions, such as an image of a blue book lying on a table includes both the book and the fact that it is blue; (3) each experience is specific and different from all other experiences; (4) each experience is a whole thing and not reducible to its parts; and (5) each experience is definite and flows at a certain speed.

According to Tononi and Koch, the key structure of a conscious thing is a set of elements that can act on each other, forward and backward, in an interconnected cause-and-effect manner, modifying and changing themselves with the interactions. The ability for elements to act on each other in both directions is critical to the theory. (A can pass information to B, and B can pass information to A.) Such two-way interactions in both directions contrast strongly with systems in which elements act only in one direction, called "feed forward networks." An example of a feed forward network is the parlor game in which a group of people form a line; the first person in line whispers something to the second person, who then whispers what she heard to the third person, who whispers what she heard to the fourth person, and so on. Information flows only in one direction. For consciousness, according to IIT, every part of the system must be able to affect and be affected by every other part of the system. Tononi and Koch go further and develop a quantitative measure of consciousness, denoted by Φ, which is related to the number of interacting elements and the number of cause-and-effect connections between them. According to this measure, the cause-and-effect structure of a highly conscious system would be much reduced if the system was subdivided.

An interesting question is the relationship between consciousness, as defined by Tononi and Koch, and life. Biologists define a living thing as an entity that has some kind of membrane separating it from the external world, that can reproduce, that can utilize energy sources, and that can evolve. Of course, there is a certain arbitrariness to these characteristics of "life." Viruses have all of these characteristics except the ability to reproduce on their own. In the future, we may find other such entities that have some but not all of these characteristics. The line between life and nonlife may not be so sharp. According to Tononi and Koch's view, a thing could be conscious without being alive in the biological sense, such as an advanced computer that could act and make changes in itself, communicate with the outside world, but could not utilize energy sources on its own and thus had to be plugged into a wall outlet. On the other hand, there are entities we would consider alive without being conscious, such as people in comas. As Koch said to me, "There is a two-way disassociation between consciousness and life. We know from patients in a coma or in a deep vegetative state, like [the American woman] Terri Schiavo [1963–2005], that they are technically not conscious, but they are alive. So, you can have life without consciousness, and consciousness without life." Deep sleep may be another example of life without consciousness.

. . .

It seems reasonable that a sufficiently complex system could have all the attributes of consciousness without being alive. However, as with a human being or a dolphin, another conscious being might never know what it *feels* like to be that particular system. The first-person/third-person divide might never be bridged. (Personally, I would love to know what it *feels* like to be a computer.) Regardless, we have shown that the behavioral manifestations of consciousness can be directly associated with the material brain. Further, we have found some of those manifestations in nonhuman animals and even suggested an evolutionary pathway to get from unconscious organisms to human brains.

Given that we cannot cross the first-person/third-person divide, that we cannot express in any objective way what it feels like to be a conscious being or thing, can we at least understand how a congregation of billions of material neurons might produce something as complex and qualitatively novel as consciousness? The answer is yes, and it comes from a study of what are called emergent phenomena. Emergence or emergentism is collective behavior of a complex system with many parts that is not apparent and often not predictable by understanding the individual parts. The modern understanding of emergentism can be traced back to the British philosopher John Stewart Mill (1806–1873), who said, essentially, that a complex system can be greater than the sum of its parts. Mill used the example of water, in which the chemical combination of oxygen and hydrogen produces a third substance whose properties are totally different from either of the two substances that created it.

The brains of crows, dolphins, and humans, with billions of neurons and trillions of connections between them, are more complex than any other natural phenomena we know of. It has been estimated that the human brain can store about 2.5 million gigabytes of information, about fifteen times as much data as the largest computer built on Earth (as of 2021). But it is not just the number of neurons that contributes to the brain's complexity. Each neuron connects to thousands of other neurons, and it is this immense network of connections that leads to spectacular emergent phenomena. At the beginning of his book *The Quest for Consciousness,* Koch writes, "The abilities of coalitions of neurons to learn from interactions with the environment and from their own internal activities are routinely underestimated. Individual neurons themselves are complex entities with unique morphologies and thousands of inputs and outputs . . . Humans have little experience with such a vast organization."

To better understand emergent phenomena in general, here are some examples.

Protein folding: Proteins interact with other metabolic molecules through electrical forces, and those forces in turn depend on the protein's structure in three-dimensional space. A protein consists of hundreds to thousands of building blocks called amino acids, which

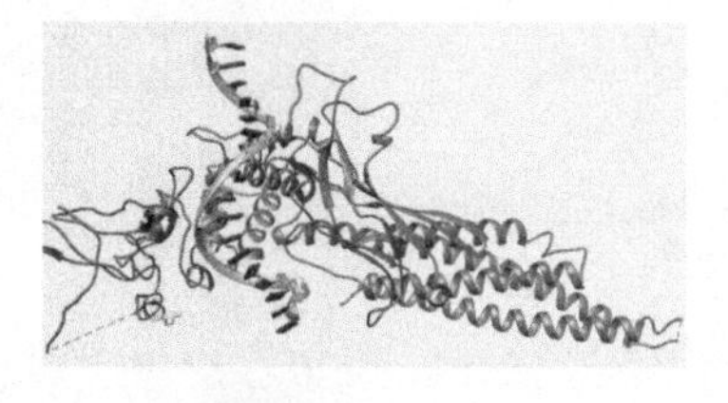

are initially created in a particular sequence along a one-dimensional line. As the protein is being manufactured, the electrical forces of the thousands of its pieces, the amino acids, work together to twist and fold the protein into its 3D shape. The final structure depends also on the molecular environment of the protein. The above illustration is a protein named STAT3, which is made up of about seven hundred and seventy amino acids. Every STAT3 protein has exactly this complex shape. Although we understand the structure of each of the individual amino acids in STAT3, the complex shape of the folded protein is far from evident from its parts. Proteins that are folded incorrectly, due to genetic defects or other errors in the sequence of amino acids, do not function properly and can cause illness and death.

Snowflake design: Snowflakes come in an enormous diversity of patterns, although all have a six-sided symmetry. It is believed that that symmetry derives from the angle at which the hydrogen atoms protrude from the oxygen, being about 120 degrees. But the particular shape of each snowflake is determined in a complex way as the emerging snowflake falls through the atmosphere and experiences random and fluctuating changes in temperature and pressure. The combined system of water molecules and air molecules is far too complex to predict the final shape and structure of any particular snowflake.

Termite cathedrals: Termites in colonies are known to build large and complex mounds, known as cathedrals. The cathedrals sometimes have elaborate galleries and chimneys to control air flow, temperature, and humidity. To build such a complex structure, there would seem to be some kind of master plan, executed by the hundreds of thousands of termites in the colony. But individual termites, which are blind, cannot perceive even the overall shape of a mound, much less direct its design. Somehow, the complex mound arises from the collective behavior of the full colony. Researchers believe that termites exchange chemical signals with one another and also respond to cues from airflow and temperature that are affected by the shape of the mound.

Neuroscience suggests to us that the emergence of consciousness in advanced brains such as the human brain, while being enormously more complex than the emergent phenomena illustrated above, is not different in kind. In particular, consciousness can emerge from the collective interaction of billions of neurons, following known laws of chemistry, physics, and biology, without the intervention of some additional ethereal or psychic force.

In summary, we have proof of a material system—the brain—that can receive visual, auditory, and other sensory input from the outside world; has internal maps of the outside world in space and time; has the ability to store information from past inputs (memory); has a vast hierarchy of agents (neurons) that can rapidly communicate with one another; is more complex than the largest computer we have built and hugely more complex than the cooperative agents that produce protein folding, snowflakes, and termite cathedrals. Even an intelligent being from another world, with a very different kind of brain, would know that such a system is capable of spectacular and novel phenomena. Consciousness, evidently, is such a phenomenon. In the next chapter, I will suggest that spirituality is as well.

Near the end of my conversation with Professor Koch, I described an experience I had some years ago. I was out on the ocean alone in a small boat, late at night, coming back to my home on a small island. It was a clear night, and the sky bristled with stars. No sound could be heard, except for the soft drone of my engine. Taking a chance, I turned off the engine. It got even quieter. Then I lay down in the boat and looked up. After a few minutes, my world dissolved into the star-littered sky. The boat disappeared. My body disappeared. Awareness of my self and my ego disappeared. And I found myself falling into infinity. I felt an overwhelming connection to the stars, as if I were part of them. And the vast expanse of time—extending from the far distant past long before I was born and then into the far distant future long after I would die—seemed compressed to a dot. I felt connected not only to the stars but to all of nature, and to the entire cosmos. I felt part of something much larger than myself. After a time, I sat up and started the engine again. I had no idea how long I'd been lying there looking up.

I asked Professor Koch whether he thought such an experience could arise from mere atoms and molecules. "First, it is a true experience," he said. "I call those mystical experiences. You can get them in near death experiences, you can get them with a drug called 5-MeO-DMT, you can get them when you meditate. We know that our brain can produce love and hate. This is another feeling that the brain can have. And experience shows that our brain can produce all these feelings of love and hate, of ecstasy, of feeling connected."

4

To See a World in a Grain of Sand

From Consciousness to Spirituality

One morning in Maine, soon after dawn, I stood by the ocean just as a light fog began moving in. The rising Sun became a gauzy fire. Suddenly, the air started to glow. Fog scattered the sunlight, bounced it around and back and forth until each cupful of air shone with its own source of light. In all directions, the air beamed and shimmered and glowed, and the gulls stopped their squawking and the ospreys became quiet. For some time, I stood there spellbound by the silence and the glowing air. I felt as if inside a cathedral of sunlight and air. Then the fog burned away and the glow disappeared.

Hinduism has a concept called *darshan,* which is the opportunity to experience the sacred. One is advised to be open to such experiences.

In this chapter, I'll suggest that spirituality follows naturally from a material brain—through the path of consciousness, high intelligence, and the evolutionary forces that shaped *Homo sapiens*. In understanding the origins of spirituality in this way, I do not mean to diminish such a majestic and profound feeling. Spiritual experiences are among the most memorable moments of our lives. I would like to suggest that they are as natural as hunger or love or desire, given a brain of sufficient complexity.

Those who believe in an omniscient and purposeful Being that created the universe often associate spirituality with that Being. One of the most beautiful and compelling statements of that association can be found in William James's landmark book *The Varieties of Religious Experience* (1902), in which a Christian clergyman describes an immediate and vital transcendent experience:

> I remember the night, and almost the very spot on the hilltop, where my soul opened out, as it were, into the Infinite, and there was a rushing together of two worlds, the inner and the outer. It was deep calling unto deep—the deep that my own struggle had opened up within being answered by the unfathomable deep without, reaching beyond the stars. I stood alone with Him who had made me, and all the beauty of the world, and love, and sorrow, and even temptation. I did not seek Him, but felt the perfect unison of my spirit with His.

The clergyman clearly attributes to God his profound feelings of connection to the cosmos. A similar cosmic connection is described by India's great Hindu poet Rabindranath Tagore (1861–1941), in his *Gitanjali:*

> Thou [God] hast made me endless,
> such is thy pleasure . . .
> The same stream of life that runs
> through my veins night and day runs
> through the world and dances in
> rhythmic measures. It is the same life
> that shoots in joy through the dust of
> the earth in numberless blades of grass
> and breaks into tumultuous waves of
> leaves and flowers.

In Islam, we have a record of Muhammad's first spiritual revelation, as recorded by his earliest known biographer, Ibn Ishaq (704–767):

> When I was midway on the mountain, I heard a voice from heaven saying, "O Muhammad! thou art the apostle of God and I am Gabriel." I raised my head towards heaven to see (who was speaking), and lo, Gabriel in the form of a man with feet astride the horizon . . . I stood gazing at him . . . moving neither forward nor backward; then I began to turn my face away from him, but

> towards whatever region of the sky I looked, I saw him as before.

Of course, one of the best-known spiritual experiences and its association with God is the Old Testament account of Moses and the burning bush:

> And the angel of the LORD appeared unto him [Moses] in a flame of fire out of the midst of a bush; and he looked, and, behold, the bush burned with fire, and the bush was not consumed.

The experiences described here, which express many of the features of spirituality as I have defined it—feelings of connection to nature, the cosmos, and other people; the feeling of being part of something larger than one's self, the appreciation of beauty; the experience of awe—are *religious* experiences, in that they are all mediated by an omniscient and purposeful Being and Creator we call God. As such, the experiences may be considered to derive from God, or from the God-like soul in us, or from the pervasiveness of God in nature and the cosmos. For the unnamed clergyman and for Tagore and Muhammad and Moses, the existence and spiritual power of God are implicitly assumed.

I respect these beliefs and their divine attributions. My aim here is to show that the same spiritual feelings can arise completely from the forces of Darwinian natural selection and the capacities of a highly intelligent brain, without other agency. In other words, I am discussing here a nonreligious spirituality and its evolutionary origins, although the feelings of connection and awe may be quite similar to those of religious spirituality.

When I talk about the forces of natural selection, I do not necessarily mean that every element of spirituality, as I have defined it, has direct survival benefit. In 1979, evolutionary biologists Stephen Jay Gould and Richard Lewontin coined the word "spandrel" to mean animal traits that were not adaptive in themselves but rather by-products of other traits that did have survival benefit. For example, eye color and earlobe size are not traits with any particular survival value, but bodily color and ears clearly have survival benefit. The ability to write poetry does not have any evident evolutionary advantage, but such ability may be the by-product of a sensitivity to sounds and rhythms, which probably did have survival benefit.

My thesis is that spirituality is a spandrel. The desire for connection and belonging, to nature and to other people; the feeling of being part of something larger than ourselves; the appreciation of beauty; the experience of awe; and the creative transcendent experience—all, I claim, are by-products of other traits that had evolutionary benefit. The first four of these need little explanation. The creative transcendent experience is a name I give to that exhilarating, soaring sensation when we produce something new in the world, discover something new, find ourselves in a state of pure seeing. Painters, musicians, dancers, novelists, scientists, and all of us have experienced the creative transcendent.

Some of the examples of spirituality I have given, such as my communion with juvenile ospreys and with the stars on a clear night, are particular transcendent experiences taking place at a particular time and place. In fact, much of our joy in being in the world derives from such particular experiences. The collection of all such moments forms the edifice of spirituality. A fascinating paradox is that most transcendent experiences are completely ego-free. In the moment, we lose track of time and space, we lose track of our bodies, we lose track of our selves. We dissolve. And yet, as I suggest, spirituality emerges from consciousness and the material brain. And the paramount signature of consciousness is a sense of self, an "I-ness" distinct from the rest of the cosmos. Thus, curiously, a thing centered on self creates a thing absent of self.

I suggest that the driving forces for the emergence of spirituality are both biological and psychological: a primal affinity for nature, a fundamental need for cooperation, and a means of coping with the knowledge of our impending death. Some of these forces can be found in nonhuman animals, of course, but the full experience of spirituality may require the higher intelligence of *Homo sapiens*.

I will now consider, one by one, the various elements of spirituality and their origins.

. . .

In his famous essay "On Nature," Ralph Waldo Emerson (1803–1882) expresses the oneness of all things in nature, including us human beings: "So poor is nature with all her craft, that, from the beginning to the end of the universe, she has but one stuff—but one stuff with its two ends, to serve up all her dream-like variety. Compound it how she will, star, sand, fire, water, tree, man, it is still one stuff, and betrays the same properties."

That oneness, even the dissolution of human self into nature, is beautifully expressed by the late Mary Oliver in her poem "Sleeping in the Forest" (1978):

> I thought the earth
> remembered me, she
> took me back so tenderly, arranging
> her dark skirts, her pockets
> full of lichens and seeds. I slept
> as never before, a stone
> on the riverbed, nothing
> between me and the white fire of the stars
> but my thoughts, and they floated
> light as moths among the branches
> of the perfect trees. All night
> I heard the small kingdoms breathing
> around me, the insects, and the birds
> who do their work in the darkness. All night
> I rose and fell, as if in water, grappling
> with a luminous doom. By morning

I had vanished at least a dozen times
into something better.

We humans (of the genus *Homo*) have spent most of our evolutionary history in a natural environment—lakes, oceans, trees, soil, grass, birds, mountains, sky. In quantitative terms, we've lived close to the land for about one hundred thousand human generations. Thus, there was almost certainly survival benefit in an attentiveness to nature. The brick and steel buildings we now inhabit for most of our daily hours are a very recent development in our two-million-year history. When we are in nature, communing with ospreys or looking up at the stars on a clear summer night, we get back in touch with something deep inside of us—hardwired into our brains. The late distinguished biologist and naturalist E. O. Wilson has used the word "biophilia" to mean "the innate tendency to focus on life and lifelike processes." (The term was first coined in 1964 by the social psychologist Erich Fromm to refer to our attraction to living things.) Wilson says that "the crucial first step to survival in all organisms is habitat selection. If you get to the right place, everything else is likely to be easier. Prey become familiar and vulnerable, shelters can be put together quickly, and predators are tricked and beaten consistently. A great many of the complex structures in the sense organs and brain that characterize each species serve the primary function of habitat selection."

The deep sensitivity to the sounds, sights, and smells of nature, formed hundreds of thousands of years in our past, must still be part of our DNA today, which itself has been cooked and recooked over the history of life. So many of our primal instincts—such as the urge to protect ourselves in the face of the unknown, the overpowering desire to love and care for our children, sexual attraction—were born from survival strategies and the forces of natural selection. It seems plausible that an affinity for nature would have arisen from a similar origin, now buried in our psyche.

Although it is difficult to prove cause-and-effect relationships for events that happened hundreds of thousands of years ago, our modern evolutionary biologists have been able to perform experiments to explore the causal interplay between evolution and environment, since the DNA of some animals and plants evolves quickly, during the lifetime of a single experiment. For example, David Reznick, an evolutionary biologist at the University of California, Riverside, and colleagues have shown that tropical fish called guppies produce more babies and develop new escape abilities and body shapes when exposed to a highly predatory environment. In other experiments, Reznick and colleagues cultivated two different populations of guppies, one from an environment with a high density of predators and a second with a low density of predators. The two different populations had very different impacts on an ecosystem consisting of algae and insect larvae. The first ate mostly the insect larvae, while the second ate mostly the algae. After only four weeks, the ecosystem surrounding the first population had evolved to contain lots of algae and little insect larvae, with the reverse for the second population. The two populations also created differences in the rate of recycling of nutrients, such as nitrogen and phosphorus.

Such studies demonstrate the unsurprising conclusion that living organisms and their ecosystems evolve and adapt to each other. Successful early humans would have been those who could adapt to their natural environment, and an attentiveness to that environment would have enhanced the adaptation.

In 2004, the social psychologists Stephan Mayer and Cindy McPherson Frantz, at Oberlin College, developed something called the "Connectedness to Nature Scale" (CNS), a set of statements that could be used to measure a person's feeling of affinity for nature. After respondents answered "strongly disagree," "disagree," "neutral," "agree," or "strongly agree" to each statement, an overall score could be computed for each participant. A few of the fourteen statements of the CNS test are:

> I often feel a sense of oneness with the natural world around me.
>
> I think of the natural world as a community to which I belong.
>
> When I think of my life, I imagine myself to be part of a larger cyclical process of living.

I feel as though I belong to the Earth as equally
as it belongs to me.
I feel that all inhabitants of Earth, human, and
nonhuman, share a common "life force."

Since 2004, psychologists have undertaken a number of studies to investigate correlations between scores on the Mayer-Frantz CNS test and previously well-developed methods of measuring happiness and well-being. In 2014, psychologist Colin Capaldi and colleagues did a meta-analysis of such correlations, combining thirty previous studies involving more than eighty-five hundred participants. The psychologists found a significant correlation between nature connectedness and life satisfaction and happiness. Associations were the strongest between happiness and the inclusion of nature in the understanding of oneself. The psychologists write that "individuals higher in nature connectedness tend to be more conscientious, extraverted, agreeable, and open . . . Nature connectedness has also been correlated with emotional and psychological well-being." Such conclusions remind us that urges, instincts, desires, and affinities formed in our development a million years in the past are still present in our psyches today.

Professor Frantz herself remembers the strong impact of nature in her childhood growing up in New Jersey, where her woodsy backyard contained a big hill and lots of rocks and where she would spend hours creating an imaginary world. "If we are more in tune with our natural environment," she told me, "we are likely to respond more effectively to cues, to changes in the circumstances. If we depend on an ecosystem, we need that ecosystem to be stable and healthy." Her words echo the statements of the late E. O. Wilson.

Just as evolutionary forces probably shaped our feelings of a deep connection to nature, they also probably shaped our need for connection to other human beings, which, in turn, is related to our feelings of being part of something larger than ourselves.

In early hunter-gatherer groups, occupying at least 90 percent of human history, members of the group would have been highly dependent on one another for survival. Danger was always nearby. The hunters went out for food, while other adults protected the children, kept the fire going, and fortified the shelter in communal settings. Being shunned or separated from the group probably would have brought a quick death. Professor Frantz says that there are definite psychological similarities between the kinds of relationships we have with nature and those we have with people. She told me that "one of the adaptive strategies that humans have is that we live in these cooperative social groups. For our ancestors, to not be a member of the group would have meant

a dramatically higher chance of dying and not passing on their genes . . . We evolved these core social motives because they helped keep people alive. The most powerful of those is the need to belong."

Stuart West, professor of evolutionary biology at Oxford University, also underscores the need for cooperation in early human living groups. He points out that cooperation is a potentially costly behavior that directly benefits others more than oneself, yet it is prevalent throughout the animal kingdom. He gives two explanations. First is reciprocity. People are more likely to help others who have helped them. Secondly, "cooperation is directed toward related individuals, who share the same genes. By helping a close relative reproduce, an individual is still passing copies of [his/her] genes onto the next generation, just doing so indirectly." From archaeological evidence in such places as the Neanderthal and

Cro-Magnon caves of France, we know that early cave dwellers lived in small groups of perhaps twenty people. Most members of such a group would have been family members and close relatives. I suggest that the need to belong to a group, with clear survival benefit, is related to the desire and feeling of being part of something larger than oneself. Frantz says that both our need to connect with nature and our need to connect with other people "stem from an individual's tendency to define ourselves as embedded in something larger than the self. This embeddedness is not only a more accurate understanding of humans' physical and psychological reality, but also brings clear mental health benefits."

Boston area psychiatrist W. Nicholson Browning sees the need to belong from the opposite end: the negative effects of social isolation. "One of the most challenging experiences for human beings is aloneness," he told me. He went on to say:

> Our horror of aloneness is fundamental to our desire for connection to the world. There is a compelling expression of the theme in the old Kubrick movie *2001, A Space Odyssey*. The astronauts, traveling in deep space, are awakened from their suspended animation by the computer, HAL. They are metaphorically born out of tiny capsules to begin functioning on the mother ship. One goes outside the ship in a tiny pod and then returns to reenter the ship by telling the computer to "open

> the pod bay doors, HAL." But the computer, having come to the conclusion that it is more capable than the humans, refuses and sends the astronaut tumbling off into the void of space to evaporate there like a tiny drop of water. I have asked perhaps sixty or seventy people if, having seen the movie, they remember that scene, and I think every person has said yes, quite vividly. I think that (unscientific) sample of people all felt profoundly the horror of, not simply the death of the astronaut, but the horror of being lost in the utter emptiness and isolation of the void. From my own personal experiences of being with people dying, there is often a wish to leave the suffering physical body, but a great sorrow at leaving our connection to the people in our lives.

Psychologists have devised a test called the "UCLA Loneliness Scale," analogous to the "Connectedness to Nature Scale," to measure the degree of loneliness in an individual. The instrument has such questions as: "I am unhappy doing so many things alone." "I have nobody to talk to." "I cannot tolerate being so alone." "I lack companionship." "I feel as if nobody really understands me." In a study of 240 people ages forty-seven to fifty-nine years old, Andrew Steptoe, a psychologist and epidemiologist and head of the Department of Behavioural Science and Health at University College London, and his colleagues have found that people with high measures on the loneliness scale responded much more negatively to stress than less lonely people. The difference actually shows up in the physiology of the body. Steptoe et al. found that lonely individuals, when exposed to stress, had higher levels of fibrinogen, natural killer cells, and cortisone. Fibrinogen is a blood protein involved with clotting. High levels can cause blood clots harmful to the brain. Natural killer cells are part of the immune system. An especially high level compared to a control group means that the body has overreacted to stress. A hyperactive immune system can cause bodily damage on its own. Thus, there are physiological manifestations of the mental state of loneliness and lack of connectedness to other human beings. The need for connection to other human beings and the consequences of not having that connection are evidently built into the biology and chemistry of our bodies.

A number of researchers have suggested that in our evolutionary history the need for social attachments and belongingness may have become associated with the physical pain system, "borrowing the pain signal to warn against social disconnection." A fascinating and compelling piece of this hypothesis is that commonly used words for social rejection, such as "hurt," "crushed," "cut," and "slapped," are the same as used for physical pain. Geoff MacDonald, professor of psychology at the University of Toronto, and Mark Leary, professor of psychology and neuroscience at Duke University, have shown that the similarity between words for social pain and for physical pain is present in languages all over the world.

The need to connect to other human beings is manifested quite literally by our need to be physically touched by others. This need is apparent in a variety of animals. Starting in the 1950s, American psychologist Harry Harlow and his collaborators showed that rhesus monkeys raised in isolation, unable to have physical contact with their mothers, showed disturbing behaviors, such as staring ahead blankly, circling their cages, and engaging in self-mutilation. (These days, such experiments would be strongly condemned by the American Society for the Prevention of Cruelty to Animals.)

A procedure for the care of preterm human babies called kangaroo care involves skin-to-skin holding of the baby against the mother's or father's bare chest. A study by pediatrician Ruth Feldman and her colleagues at Bar-Ilan University in Ramat Gan, Israel, compared preterm babies who received kangaroo care in the neonatal unit to preterm babies who received standard care. The study found that after thirty-seven weeks, the infants who experienced kangaroo care showed more alertness, more motor control, and less gaze aversion than the infants in the control group.

It seems clear that physical touching is important for the normal development of the higher animals, including humans. Such touching, in turn, is almost certainly related to a need to be connected to the larger world.

. . .

A detachment from one's Self plays an important role in various aspects of spirituality, especially the transcendent experience of being connected to something larger than ourselves. When we open ourselves up to a larger world, we are in some ways subjugating and dissolving our individual egos. For a few moments at least, we let go of Self. Thus, there would seem to be an anticorrelation between the degree to which we focus on our individual selves and the degree to which we can connect to things larger than ourselves. More self, less connection to the larger world.

Professor Frantz and her colleagues have provided support for such a hypothesis. In a paper titled "There is no 'I' in nature: The influence of self-awareness on connectedness to nature," published in the *Journal of Environmental Psychology,* the researchers report that in a study of some sixty participants, heightened self-awareness is associated with *less* connectedness to nature. Self-awareness here is quite literal: the extent to which an individual sees herself to stand out from the background of the rest of the world, "a separate and discrete object in the world." Frantz and colleagues overtly manipulated their subjects' self-awareness by seating them facing either the reflective (high self-awareness) side of a mirror or the nonreflective (low self-awareness) side. The participants then answered questions on the CNS test to measure their feeling of connectedness to nature and the larger world. Although such experiments seem simplistic, their results are consistent with the commonly felt loss of one's self during transcendent experiences.

It is worth pausing here to acknowledge that the relative importance of one's self is partly shaped by one's society. Cultural anthropologists, sociologists, and psychologists have long noticed substantial differences between Westerners and Easterners in their attitudes toward the individual versus the larger group. Westerners (Americans, Europeans, Australians, etc.) celebrate the individual and the individual's freedom, independence, and autonomy, while Easterners (Chinese, Japanese, Koreans, etc.) prioritize the group over the individual and emphasize the interdependent relationships between members of the group. The two psychological and cultural dichotomies are referred to as "individualism" versus "collectivism." One particularly interesting feature of this dichotomy: To study a phenomenon or analyze a problem, individualists break the thing down into its component parts. Collectivists experience phenomena holistically, and in terms of the relationships between all the parts.

Some years ago, I visited the Foreign Correspondents' Club of Japan, in Tokyo. While trading stories over drinks, the journalists exchanged business cards. As I remember, the cards of the Japanese journalists had the name of their publication in large letters at the center of the card and their personal name in small letters in a corner. It was the reverse for the Western journalists.

Collectivism must have come first in our evolutionary history. Our earliest ancestors, living together in small communes, had to be collectivists in order to survive. The group was paramount. If a member of the tribe became too independent of the group, she would die. Thus, in the long history of *Homo sapiens,* individualism is a relatively recent phenomenon. How did it become more prominent in the West than in the East? The question fascinates me. There are probably many historical, cultural, and psychological factors in such a development.

Early philosophers reflected the individualism/collectivism dichotomy. According to the ancient Greek writer Pausanias, the well-known adage "Know thyself," expressing the power and responsibility of the individual, was inscribed on the wall of the Temple of Apollo at Delphi. The importance of self-knowledge of the individual is repeated by Socrates in Plato's *Apology:* "The unexamined life is not worth living." Such were among the roots of Western philosophy.

Compare these statements centered on the individual to the words of Zeng Shen (505 BC–436 BC), a disciple of Confucius: "In planning for others, have I been loyal? In company with friends, have I been trustworthy? And have I practiced what has been passed on to me?"

A possible factor in the origins of individualism is the role of exploration. Historian Frederick Jackson Turner, in his highly influential essay "The Significance of the Frontier in American History" (1893), argued that the process of exploration of the American West helped shape the individualism and independence of the American character. When we explore, we leave the comfort and safety of the larger community and go out on our own. Turner's frontier thesis has been given support in a study by social psychologist Shinobu Kitayama of the University of Michigan and colleagues. These researchers studied the values and psychology of Japanese people living in the northern territory of Hokkaido, which in the late nineteenth century experienced a rapid migration and exploration akin to the pioneering settlement of the "Wild West" in America. Kitayama and colleagues found that the people of Hokkaido were happiest as a result of their *individual* accomplishments, while the people of mainland Japan reported being happiest when connected to their surrounding society. Furthermore, the Hokkaidoans placed more value on personal independence than other Japanese.

Considering these cultural differences and their origins, one can conclude that Westerners may have more inertia to overcome than Easterners in letting go of the individual ego and opening up to the larger world. That is not necessarily to say that Easterners are more spiritual in general than Westerners. But the relative priorities of the individual versus the group are certainly different, and those differences must affect one's relationship to the world beyond oneself.

An unexpected statement of our common human bonds comes from Albert Einstein. The great physicist was an inadequate family man and a loner for most of his life, but he did have this to say about human connections in an essay he wrote for *Forum and Century* in 1931:

> How strange is the lot of us mortals! Each of us is here for a brief sojourn; for what purpose he knows not, though he sometimes thinks he senses it. But without deeper reflection one knows from daily life that one exists for other people—first of all for those upon whose smiles and well-being our own happiness is wholly dependent, and then for the many, unknown to us, to whose destinies we are bound by the ties of sympathy.

Einstein's mention of our "brief sojourn" points to what I consider a major driving force behind our urge to connect with other people and with the larger cosmos: our personal mortality and our desire to transcend the limitations of our physical bodies. Of course, the awareness of death can be found in nonhuman animals, as discussed for the group of chimpanzees in chapter 3. But the longing to become part of existence on a cosmic scale requires an additional sophistication and intelligence, a sense of the great chain of human lives stretching back hundreds of thousands of years in the past and an untold number of years in the future, all of us connected through the parents of parents and the children of children.

Such a feeling and need for connection on the cosmic scale is augmented by a comprehension of our place on Earth, awe of the night sky, and other sophisticated understandings. For thousands of years, the stars were considered indestructible and eternal. In the incantation for Unis, quoted in chapter 1 and dating back to 2315 BC, the deceased pharaoh is beckoned to join the "imperishable stars." Two thousand years later, Plato chose the stars to be the final destination of all moral human beings after their fleeting time on Earth: "And having made [the universe], the Creator divided the whole mixture into souls equal in number to the stars, and assigned each soul to a star . . . He who lived well during his appointed time was to return and dwell in his native star." Two thousand years after Plato, we have Sigmund Freud's discussion of an "oceanic feeling." In his *Civilization and Its Discontents,* Freud endorses the sentiments of French novelist and dramatist Romain Rolland, winner of the 1915 Nobel Prize in Literature. In a letter to Freud, Rolland suggested that the source of religious energy lies in an "oceanic feeling," which is a "sensation of 'eternity,' a feeling as of something limitless, unbounded—as it were, 'oceanic' . . . a feeling of an indissoluble bond, of being at one with the external world as a whole."

As mentioned earlier, the American cultural anthropologist Ernest Becker has argued that our entire civilization is a "defense against death." In the preface of his Pulitzer Prize–winning book *The Denial of Death,* Becker writes, "the idea of death, the fear of it, haunts the human animal like nothing else; it is a mainspring of human activity—activity designed largely to avoid the fatality of death, to overcome it by denying in some way that it is the final destiny for man."

Indirectly, these yearnings relate to the question of whether we are alone in the universe. On March 6, 2009, the Kepler scientific observatory, named after the Renaissance astronomer Johannes Kepler, was launched into space, specifically designed to search for planets outside our solar system that would be habitable—that is, neither so near their central star that water would be boiled off, nor so far away that water would freeze. Most biologists consider that liquid water is a precondition for life, even life very different from that on Earth. Kepler has surveyed about 150,000 Sun-like stellar systems in our galaxy and discovered over twenty-six hundred alien planets. Although the satellite stopped functioning in 2018, its mountainous stockpile of data is still being analyzed. For centuries, we human begins have speculated on the possible existence and prevalence of life elsewhere in the universe. For the first time in history, we can begin to answer that profound question, Are we alone in the universe?

I suggest that among the many motivations behind the conception and creation of Kepler has been the desire to connect with the rest of the cosmos, to find other living and thinking beings that share the spectacle of this wondrous universe we find ourselves in. Our individual lifetimes are but brief flares in the unfolding history of the cosmos. A discovery of other living things on other worlds would reveal a larger tapestry of which we are a part.

I would also suggest that a psychological force behind the entire enterprise of science—although that force may operate at the subconscious level—is the desire to find truths that endure beyond our individual human lives. Newton's laws of motion will last for millennia. As will Darwin's natural selection. Physicist Kip Thorne (winner of the 2017 Nobel Prize for his work on the detection of gravitational waves) recently explained to me part of his personal motivation as a scientist: "When we look back on the era of the Renaissance and ask what is the legacy that our ancestors of that era left for us, most of us respond: great art, great architecture, great music, and the scientific method. Similarly, when our descendants several centuries from now ask the same question about our legacy to them, I think a big part of their answer will be an understanding of the universe and the physical laws that govern it."

My own work in science has not been nearly so significant as that of Thorne, but I can remember my deep satisfaction on the few occasions when I have discovered something new about the physical world. The neighbor's dog barking across the street, my headache and heaviness from staying up all night, the spilled tea dripping on the floor, the fresh bruise on my leg from bumping into a chair all seemed like a gust of wind compared to the equations I had just written down for the behavior of a hot gas or a cluster of stars orbiting a black hole.

In Christian theology, there is a concept called "the Great Chain of Being," which refers to a hierarchy of beings on Earth and in heaven. At the top of the ladder is God. Next in line are the angels, then humans, then nonhuman animals, plants, and finally nonliving material at the bottom. The concept traces back to Aristotle's *scala naturae* ("ladder of life"). The theological Great Chain of Being establishes an encompassing framework for everything in existence. I would like to suggest a related phrase: the "Great Chain of Connection." Rather than expressing a vertical hierarchy, this phrase refers to a more horizontal network: our feeling of connection to other human beings, to nature, and to the cosmos as a whole—a feeling of being part of something much larger than ourselves. Confronted with our impending demise as individuals, is it not comforting to feel part of something larger, something that outlasts our individual lifetimes? Perhaps only the genus *Homo* is aware of its mortality; certainly such awareness requires a higher level of intelligence.

The Great Chain of Connection is not unrelated to another concept I introduced in my book *Probable Impossibilities,* a concept I call "cosmic biocentrism." That idea refers to the kinship of all living things in the universe, a kinship underscored by the recent scientific understanding that life can exist in the universe only during a relatively limited period of time. Before the onset of the era of life, the complex atoms needed for life had not yet been manufactured in stars.

After the era of life, stars will have burned out and all other energy sources available to support life will have either exhausted themselves or become unavailable due to the end of contact between galaxies. Furthermore, the fraction of material in the universe in living form is extremely small—one-billionth of one-billionth—equivalent to a few grains of sand in the Gobi Desert. For all of these reasons, life is precious, both in time and in space.

Coppery clouds. The winding swirl of a seashell. The splaying of hues in a rainbow. The reflection of stars on the skin of a still pond at night. Much of nature we consider beautiful because we are part of nature. We grew up in nature, evolutionarily speaking. Of course, there's also a cultural component to the notion of beauty, especially when it comes to the physical beauty of people: Elongated earlobes are considered beautiful by the Masai in Kenya. For centuries, the Chinese bound the feet of girls, believing that small feet were beautiful, feminine, and a sign of refinement. But some concepts of beauty seem universal and almost certainly by-products of traits with survival benefit. The botanist and geneticist Hugo Iltis (1882–1952) wrote that "man's love for natural colors, patterns and harmonies . . . must be the result to a large degree of Darwinian natural selection through eons of mammalian and anthropoid evolutionary time."

It is not hard to argue that an appreciation of color and form and other aspects of beauty had survival benefit in their relation to sexual attraction. The primal and evolutionary force behind sexual attraction, of course, is procreation, and procreation is most successful when both partners are healthy and vigorous. Health and vigor, in turn, are associated with well-formed body shape, smooth skin, good color, striking facial features, and other aspects of bodily "beauty." In fact, the neurological reactions to beauty trigger some of the same pleasure centers in the brain as eating, sex, and drugs.

Both Darwin and Freud opined that our sense of beauty originated as a strategy to promote reproduction. In *The Descent of Man,* Darwin writes, "When we behold a male bird elaborately displaying his graceful plumes or splendid colours before the female, whilst other birds, not thus decorated, make no such display, it is impossible to doubt that she admires the beauty of her male partner." Freud was reluctant to comment on the meaning of beauty, except in its relation to sex: "Psychoanalysis has less to say about beauty than about most things. Its derivation from the realms of sexual sensation is all that seems certain . . . Beauty and attraction are first of all the attributes of a sexual object."

As I mentioned earlier, some aspects of spirituality, including the appreciation of beauty, may be byproducts of traits with survival benefit rather than traits with direct survival benefits themselves. An attraction to beauty might well have other manifestations in addition to its relation to sexual attraction. In such a way, we find appealing the colors of the western sky just after sunset, the patterns formed by constellations of stars, the rise and fall of wind through trees.

Our sensitivity to beauty, combined with our kinship with the natural world, has some surprising aesthetic manifestations and interconnections. Take the golden ratio. Biologists, architects, psychologists, and anthropologists have long noted that we find especially pleasing those rectangles whose ratio of long side to short side is approximately 3:2. That ratio is close to what is called the golden ratio. Two numbers are in the golden ratio if the ratio of the larger number to the smaller is the same as the ratio of their sum to the larger number. From this seemingly simple definition, we can determine that the golden ratio is approximately 1.61803. (See the endnotes for the exact value.)

Now, we enter the kingdom of magic. The twelfth-century Italian mathematician Leonardo Fibonacci (ca. 1170–1240) discovered an interesting sequence of numbers, called the Fibonacci series:

0, 1, 1, 2, 3, 5, 8, 13, 21, 34, 55, . . .

Each number of the sequence, after the zero, is the sum of the previous two numbers. As you can test for yourself, the ratio of a number in the sequence to the one before it approaches the golden ratio as we go to bigger and bigger numbers. For example, 21/13 = 1.615, 34/21 = 1.619, 55/34 = 1.6176. So this special series of numbers is closely related to the golden ratio. Even at this point, anyone with an appreciation for mathematics can see much beauty in the golden ratio and its relation to the Fibonacci series of numbers.

But there is much more to this natural magic. Consider a spiral constructed by the quarter circles connecting the opposite corners of a series of ever bigger squares whose sides are the numbers in the Fibonacci series, as shown in the diagram below:

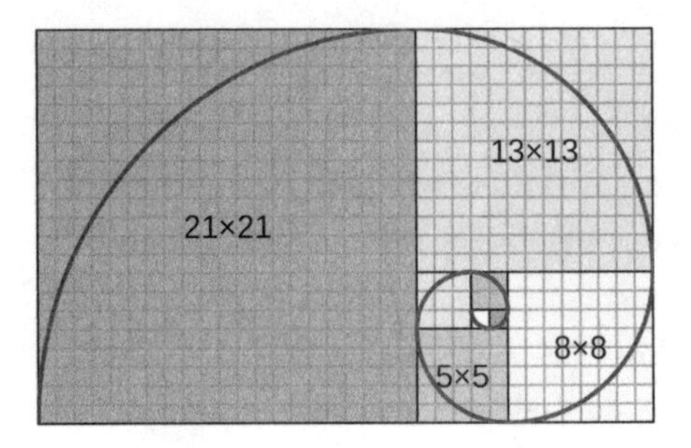

Astonishingly, many biological organisms embody this spiral. For example:

Seashell

Aloe polyphylla

With its ubiquity in nature, the golden ratio is, not surprisingly, pleasing to the human eye. Architects, ancient and modern, have built it into their constructions, sometimes unconsciously. For example, the Great Pyramid of Giza (2560 BC) has a slant height of 186.3 meters and a half base length of 115.2 meters, with a ratio of 1.6172, almost exactly equal to the golden ratio.

The CN Tower in Toronto, the tallest freestanding structure in the Western Hemisphere, has an observa-

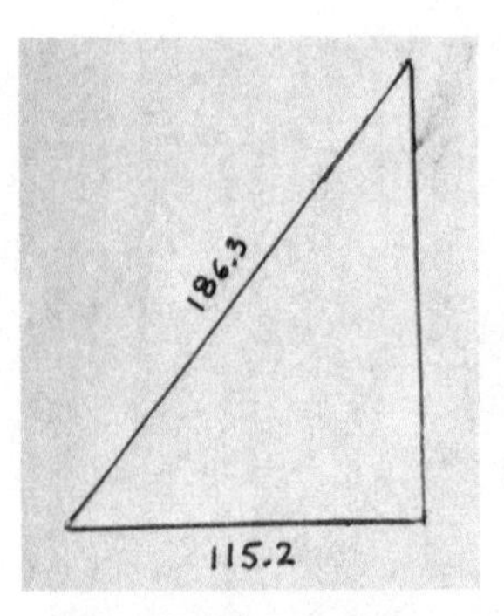

tion deck located at 342 meters and continues upward for another 211 meters after that. The ratio of these two lengths is 1.62, pretty close to the golden ratio. Evidently, the beauty of mathematics, the structures of organisms in nature, and our human sense of aesthetics all sing in unison the golden ratio.

The mechanical engineer Adrian Bejan at Duke University has offered an evolutionary explanation, based on the eye and the brain, as to why we find the golden ratio so appealing. Bejan argues that the eye and brain would have evolved to maximize the ease of flow from the visual plane to the brain. If we consider a rectangle of horizontal length l and height h, the time for the eye to scan the area of the rectangle is smallest when the eye can scan the horizontal length in the same time it takes to scan the vertical. After doing an analysis of the geometry of the eye, Bejan finds that the eye scans in the horizontal direction about 3/2 times as fast as in the vertical direction. Thus, the optimal value of l/h, which minimizes the time to scan the entire rectangle, is about 3/2, not far from the golden ratio.

It is but a short step from Bejan's analysis to argue that since many objects in nature exhibit the golden ratio in their construction, our eyes would naturally have evolved to optimize the flow of information to the brain for objects with this ratio. And a short step from there to argue why this ratio is so pleasing to the eye. The golden ratio is built into us, just as it is built into seashells and aloe plants. *Our aesthetic of beauty is literally an expression of our oneness with nature.*

Understanding these explanations of why I appreciate beautiful things, and indeed my concept of beauty, does not in the slightest diminish my pleasure and delight in gazing upon coppery clouds or spiraling seashells or the reflection of stars in the water. In fact, they enhance my pleasure, emphasizing my connections to the natural world. For me, the elegance of the mathematics of the Fibonacci series, the presence of that particular beauty in seashells and plants, and my own biological affinity for such beauty are all of a piece, a wholeness, a profound connectedness of all living things. It is all part of the Great Chain of Connection.

Many years ago, I took my then two-year-old daughter to the ocean for the first time. As I remember, we had to walk quite a distance from the parking lot to the point where the ocean slid into view. Along the way, we passed various signs of the sea: sand dunes; seashells; sunbaked crab claws; delicate piping plovers, which would run and peck, run and peck, run and peck; clumps of sea lavender growing between rocks; and an occasional empty soda can. The air smelled salty and fresh. My daughter followed a zigzagging path, squatting here and there to examine an interesting rock or shell. Then we climbed over the crest of a final sand dune. And suddenly, the ocean appeared before us, silent and huge, a turquoise skin spreading out and out until it joined with the sky. I was anxious about my daughter's reaction to her first sight of infinity. Would she be frightened, elated, indifferent? For a moment, she froze. Then she broke out in a smile.

In a paper titled "Approaching awe, a moral, spiritual, and aesthetic emotion," psychologists Dacher Keltner and Jonathan Haidt write that awe has two distinctive features: "a perceived vastness, and a need for accommodation, defined as an inability to assimilate an experience into current mental structures." The first of these features is closely related to my definition of spirituality. The second refers to witnessing something that we do not fully understand, that takes us beyond our common experiences. We are not awed by the sound of a closing door or other routine events of daily life. To "a perceived vastness" I would add a perception of being in the presence of something larger, grander, and more powerful than ourselves. In these terms, a capacity for awe might be associated with a need to belong to something larger than ourselves, and an appreciation for beauty. We can be awed by natural phenomena—as my daughter was awed by the ocean—and we can also be awed by other human beings. People who are stronger, smarter, more talented, more powerful than ourselves. We are awed by Superman, Albert Einstein and Marie Curie, Pablo Picasso, Carl Lewis and Michael Phelps, Jane Austen, Beethoven, Abraham Lincoln, Angela Merkel, Jack Ma. One could argue that to enhance survival, members of our ancestral communities would have necessarily divided into leaders and followers. Successful members of a group must acknowledge and accept its leaders. As far as nature is concerned, we are all followers.

The Chinese poet and government official of the Tang dynasty Bai Juyi (772–846) expressed his awe of the larger world after climbing to the top of Mount Xianglu near Shaoxing, Zhejiang, China:

> Up and up, the Incense-burner Peak . . .
> My hands and feet—weary with groping for
> hold.
> There came with me three of four friends,
> But two friends dared not go further.
> At last we reached the topmost crest of the Peak;
> My eyes were blinded, my soul rocked and
> reeled.
> The chasm beneath me—ten thousand feet;
> The ground I stood on, only a foot wide.

If you have not exhausted the scope of seeing
 and hearing,
How can you realize the wideness of the world?
The waters of the River looked narrow as a
 ribbon,
Peng Castle smaller than a man's fist.

I believe that the capacity for awe also includes an openness to the world. Openness, in turn, requires a certain humility. To be open to the world is to recognize that there are things in the world that we do not yet possess, things bigger than us, things that we do not yet understand (or perhaps may never understand). Many years ago, the mathematical physicist Roger Penrose (winner of the 2020 Nobel Prize) expressed to me his view of the world: "Suppose you have something in nature that you are trying to understand, and finally you can understand its mathematical implications and appreciate it. Yet there is always some deeper significance there . . . once you have put more and more of your physical world into mathematical structure, you realize how profound and mysterious this mathematical structure is. How you can get all these things out of it is very mysterious."

Returning to the Hindu concept of *darshan*, which literally means "viewing" in Sanskrit, its fuller meaning is the experience of beholding a deity or sacred object. For Penrose, mathematics is that sacred object. For Bai Juyi, it was the view from a mountaintop. For my early morning in Maine, it was the shining air. For my two-year-old daughter, it was the ocean. The experience of *darshan* is considered to be reciprocal. When we open ourselves to the world and pay homage to that which is larger than ourselves, we receive a blessing from the outer world. We get something back. We are ourselves enriched by the larger understanding of the cosmos and our place in it.

The last aspect of my notion of spirituality, which I have called the creative transcendent, may not have the same evolutionary roots as the others, but it could well be a by-product of the urge for exploration and discovery—finding new hunting grounds, new sources of water, new sources of food. A painting, a musical composition, a poem, a novel idea, a sudden insight about how to decorate a room—aren't they all explorations of a kind? And what exactly are we exploring? I suggest that in the creative transcendent experience, we are exploring both the outer world beyond ourselves and the inner world of our minds. We are exploring our hidden capacities. When we create, we discover new things about ourselves. We find secret doors. And, perhaps most importantly, we uncover new connections between ourselves and the rest of the cosmos.

As I mentioned earlier, the creative transcendent, like other transcendent experiences, involves a total loss of self and embodiment. According to the "emptiness" ideas of Vajrayana (Tibetan) Buddhism, the self is an illusion. And the ego is an obstacle. Indeed, in Buddhism, all suffering is attributed to excessive attachment of our ego to things we do. And that ego, or sense of self, disappears during the creative transcendent experience. During the creative moment, we have no sense of our self, our body, even time and space. We are simply in the zone. We have entered a state of pure seeing, as Bai Juyi mentions in his poem. Associated with that experience is a letting go. At least temporarily, we forget our cares, we leave behind the constant rush and heave of the world. Bodiless, we travel to some other space. I am tempted to call it an ethereal space, but, being a materialist, I believe that this other space is rooted in the material brain. That said, the material brain is capable of wondrous things. And the trip seems effortless. We do not struggle in the creative transcendent. We glide.

In 1926, the British social psychologist and educator Graham Wallas proposed that creative thinking follows a series of stages: preparation, incubation, illumination, and finally verification. In the preparation stage, the person does her homework or research in a field or art form or any other endeavor, masters the tools of the craft, and defines some problem. In the incubation stage, the person mulls over the problem in various ways, sometimes unconsciously. In the illumination stage, the person achieves a new insight or shift of perspective. And in the verification stage, the person puts the insight to the test and works out the consequences. The creative transcendent would occur during the incubation and illumination stages. Here are some personal accounts.

The first is by the distinguished mathematician Henri Poincaré (1854–1912). Mathematical creativity is an intriguing special case of the creative transcendent. Practitioners and philosophers disagree on whether mathematical truth exists out there in the world, independent of the human mind—in which case mathematicians discover what is already there, like coming upon a new ocean—or whether mathematical ideas, theorems, and functions are invented out of the mind of the mathematician. In any case, Poincaré wrote these words about one of his creative experiences: "Every day I seated myself at my work table, stayed an hour or two, tried a great number of combinations and reached no results. One evening, contrary to my custom, I drank black coffee and could not sleep. Ideas rose in crowds; I felt them collide until pairs interlocked, so to speak, making a stable combination. By the next morning I had established the existence of a class of Fuchsian functions."

Painting is a more familiar form of creative activity, requiring both disciplined training and spontaneous inspiration. Following the Wallas description, my wife spent ten years studying the craft, drawing plaster busts to master light and shadow and rounded edges; now, as a professional, she will sometimes spontaneously decide to add an accent of red in the corner of a still life to provide balance and interest. The painter Paul Ingbretson (1949–), a contemporary leader of the Boston School of

American art and past president of the Guild of Boston Artists, told me, "All my awareness [while painting] is fixed on the exercise of the search for the likeness and the beauty. I do get rewards every twenty minutes or so as I accomplish minor missions and see the whole coming into view. A bit of a high, yes. Like hitting a shot in sports . . . I am always the conscious vehicle, always submitting to bringing the beauty of the thing itself. My job is to get out of the way. And in that way, I lose myself . . . Whenever I come close to the magic, I know it is something bigger than me. Even truth is merely a vehicle for beauty. I remember the first time I saw the beauty of three colors as a unity, I thought this is too much for me—like the Holy Land, perhaps. I remembered Moses taking off his shoes at the burning bush. Like that."

Creativity in science is somewhere between discovery and invention. There is already a large body of established facts about the world that cannot be circumvented when inventing new theories, what the physicist Richard Feynman described as creating while wearing a straitjacket. Scientific creativity takes place at the boundary between the known (the accumulated facts from prior experiments) and the unknown (regions of the physical world that have not yet been explored). In his autobiography, the physicist Werner Heisenberg (1901–1976) describes the transcendent moment when he realized that his new theory of quantum mechanics would succeed in describing the hidden world of the atom. At the end of May 1925, after months of struggling with his theory, he fell ill with hay fever and took a two-week leave of absence from the University of Göttingen.

> I made straight for Heligoland, where I hoped to recover quickly in the bracing sea air . . . Apart from daily walks and long swims, there was nothing in Heligoland to distract me from my problem . . . When the first terms [in the mathematical equations] seemed to accord with the energy principle, I became rather excited, and I began to make countless arithmetical errors. As a result, it was almost three o'clock in the morning before the final result of my computations lay before me . . . At first, I was deeply alarmed. I had the feeling that, through the surface of atomic phenomena, I was looking at a strangely beautiful interior, and felt almost giddy at the thought that I now had to probe this wealth of mathematical structures nature had so generously spread out before me. I was far too excited to sleep.

In the literary arts, writers cannot fully plot out their characters' actions, or else they will not come to life. There must be an element of surprise, even for the writer. The writer must disappear, or be a fly on the wall, listening to her characters speak rather than telling them what to say. In a speech delivered to the National Society for Women's Service in early 1931, novelist Virginia Woolf gave this description of her creative process:

> A novelist's chief desire is to be as unconscious as possible . . . I want you to imagine me writing a novel in a state of trance. I want you to figure to yourselves a girl sitting with a pen in her hand, which for minutes, and indeed for hours, she never dips into the inkpot. The image that comes to my mind when I think of this girl is the image of a fisherman lying sunk in dreams on the verge of a deep lake with a rod held out over the water . . . letting her imagination sweep unchecked round every rock and cranny of the world that lies submerged in the depths of our unconscious being.

Although the above accounts come from professional scientists and artists, we all have experienced some aspects of the creative transcendent—from decorating a room to playing the piano to designing a marketing plan to giving birth to a child.

I will end with an example from my own experiences with the transcendent creative. One of my first research problems as a graduate student in physics concerned the force of gravity—the question of whether the experimentally observed fact that all objects fall with the same acceleration is sufficiently powerful to rule out a group of theories of gravity, called nonmetric theories, in competition with Einstein's theory. In the grand scheme of physics, it was not so consequential a question, but it had not been previously answered. After an initial period of study and work, I had succeeded in writing down all the equations to be solved. Then I hit a wall. I knew I'd made a mistake, because a result at the halfway point was not coming out as it should, but I could not find my error. Day after day, I checked each equation, pacing back and forth in my little windowless office, but I didn't know what I was doing wrong, what I had missed. Then one morning, I remember that it was a Sunday morning, I woke up about five a.m. and couldn't sleep. I felt extremely excited. Something was happening in my mind. I was thinking about my physics problem, and I was seeing deeply into it. And I had absolutely no sense of my body. It was an exhilarating experience, a kind of rapture—an experience completely without ego.

The closest physical analogy I've experienced to this creative moment is what sometimes happens when you are sailing a round-bottomed boat in strong wind. Normally, the hull of a boat stays down in the water and the friction of the water greatly limits the speed of the boat. But in a strong wind, every once in a while, the hull of a round-bottomed sailboat will lift out of the water, get on top of the water, and the frictional drag drops to almost zero. Then the boat leaps ahead. It feels like a gigantic hand has grabbed hold of your mast and yanked you forward. You go skimming across the water like a smooth stone. It's called planing. So I awoke on this morning planing in my mind. Something had grabbed hold of me, but there was no "me."

With these sensations surging through me, I tiptoed out of my bedroom, almost reverently, afraid to disturb whatever strange magic was going on in my head, and went to the kitchen table, where my crumpled pages of calculations lay before me. Within a few hours, I had found the error and solved my physics problem. I had discovered some small truth about the cosmos, a thing hidden until I found it. And yet the "I," or at least the ego, was absent during the discovery.

It is a cool, cloud-covered morning in June, and I am kayaking in one of my favorite coves off the coast of Maine. The sky is solid white, an infinity of nothingness. The craggy shoreline curves in and out. Beyond the wandering line where water meets land, low bushes morph into trees. In the distance, a single wooden house sits on high ground, apparently abandoned but still lovely with its mottled red roof and window boxes.

Kayaking is a meditative activity. The rhythm of stroking is like the mindful breathing of Buddhism. Dip the right paddle, then the left paddle, then the right, then the left. Slowly, slowly. My kayak glides through the water without making a sound. Right, then left, then right, then left. The shore melts into a patch of green and orange, an abstract painting. Right, left, right, left.

I will take a moment to break this spell and write down my thoughts (in a small notebook I take during marine excursions). Now, I am back in my body, back in my conscious brain. Is this the real world, and the other a world of illusion? Or perhaps the reverse. As I've come to understand, a common feature of all aspects of spirituality is a loss of self, a letting go, a willingness to embrace something outside of ourselves, a willingness to listen rather than talk, a recognition that we are small and the cosmos is large. For a moment, I stop paddling and listen. I think that I hear the soft beats of my heart. Or is it the soft clapping of waves on the shore?

5

My Atoms and Yours

Science and Spirituality in the World of Today

We live in an age of science and technology: smartphones, antibiotics and vaccines, airplanes, genetic engineering, computers, the big bang, splitting the atom, spaceships to Mars, quantum physics, self-driving cars, Ritalin and Adderall, lasers. In recent years, some of these developments—although undeniably beneficial to our advance as an enterprising species—have further polarized an already polarized society.

On one extreme is the belief that science has all the answers, not simply to landing men and women on the Moon but how to structure governments and economies, how to decide if a murderer should receive capital punishment, and many other social, moral, and even aesthetic issues. According to this way of thinking, sometimes called logical positivism and sometimes scientism, if an issue or phenomenon is not subject to scientific analysis, it's not valuable. Anything that cannot be measured, weighed, and counted is not worth counting. People in this group are accused of taking the soul out of human experience.

On the other extreme are people suspicious not necessarily of science itself but of the institutions of science and its priests. This group associates the universities and laboratories and professors of science with the elite establishment that has usurped the lives of ordinary working people. This group is part of the global populist movement of recent years. People in this group, sometimes called anti-science, are accused of dismissing facts and evidence that conflict with their beliefs. In recent years, we have seen such accusations partly validated in the denial of human-caused climate change and of the outcome of presidential elections.

Of course, many of us fall somewhere between these two extremes: science and its practitioners can indeed answer many questions, but not all questions, and not those that concern social, moral, and aesthetic issues. Most scientists accept the value and validity of human experience.

Scientists themselves have sometimes unintentionally exacerbated the views of the two extremes. The institutions of science are indeed privileged. The knowledge of science and technology has great power over our lives, yet only a small fraction of us have the technical training to master that knowledge. Scientists could do a better job at reaching out and trying to understand the anti-science camp. And the anti-science camp could do a better job at trying to understand the methods of science and the manner in which scientists acquire knowledge.

There is a related issue that this book tries to address, at least tangentially. In recent years, a group of scientists and philosophers have attempted to use scientific arguments to undermine the belief in God. These people are called "the new atheists." In September 2018, I debated the most prominent of the new atheists, Richard Dawkins, at Imperial College London. In my opening statement, I described some of my personal transcendent experiences and then went on to say that such spiritual experiences are part of the profound current of feeling and response to the world that has streamed through the human condition for thousands of years—in painting, in music, in literature, in love. You can see it in the Cro-Magnon paintings in the caves of Les Eyzies and Lascaux. You can hear it in Beethoven's *Eroica*. Transcendent experiences may not be understood quantitatively or even logically—certainly not in the manner in which a physicist can calculate the number of seconds it will take a ball to fall to the floor when dropped from a height of two meters. Following my opening remarks, Dawkins walked to the podium and said that I couldn't "out-transcendent him." Of course, he said, he had experienced such moments himself. But when it comes to religious belief, part of which surely follows from the same desires and urges behind transcendent experiences, Dawkins dismisses people of faith as "nonthinkers" and labels religion as "nonsense." Such an attitude from a leading spokesperson for the scientific establishment only increases the divide between different groups of people.

Science can never disprove the existence of God, since God might exist outside the physical universe. Nor can religion prove the existence of God, since any phenomenon or experience attributed to God might, in principle, find explanation in some nontheist cause. What I suggest here is that we can accept a scientific view of the world while at the same time embracing certain experiences that cannot be fully captured or understood by the material underpinnings of the world. This perspective may not be the preferred path for all of us. But it offers many of us a way of being in the world that affirms both science and spirituality. What we need is a balance between wanting to know how the world works—the driving force of science—and the willingness to surrender ourselves to some things that we may not fully know. As I said at the beginning of this book, we human beings are both experimenters and experiencers.

One of my favorite statements by Einstein, first published in 1931, is, "The most beautiful experience we can have is the mysterious. It is the fundamental emotion which stands at the cradle of true art and true science." What did Einstein mean by "the mysterious"? I don't think he meant the supernatural or the forever unknowable. I think he meant that magical realm between the known and the unknown, a place that provokes us and stimulates our creativity and fills us with amazement. Scientists and artists, believers and nonbelievers can stand on the precipice between the known and the unknown, without fear, without anxiety, but instead with awe and wonder at this strange and beautiful cosmos we find ourselves in.

When I was a physics major in college, I learned why the sky is blue. The reason is that as incoming sunlight, composed of a full spectrum of colors, strikes air molecules, the electrons in those molecules respond to the light in such a way that the shorter wavelengths of light, toward the blue end of the spectrum, are scattered sideways much more strongly than the longer wavelengths. When we look away from the Sun, we see only scattered light. The phenomenon is known as Rayleigh scattering, named after the first person who worked out the details in 1871, the British physicist Lord Rayleigh (John William Strutt). In his calculations, Lord Rayleigh used the equations for electromagnetism recently discovered by James Clerk Maxwell (see chapter 2). And where did those equations come from? They are required to describe the behavior of a certain energy field and its symmetries. And why should such energy fields and symmetries exist? For that, we could ask Roger Penrose, who might say it is all mathematics. And yet, adds Penrose, "how you can get all these things out of [mathematics] is very mysterious." If you pull on the thread far enough, you ultimately arrive at the mysterious. When at age twenty I learned why the sky is blue, my awe of the universe did not diminish.

I will end with a final portrait of spiritual materialism, as I see it. There is very good scientific evidence that all of the atoms in our bodies, except for hydrogen and helium, the two smallest atoms, were manufactured at the centers of stars. If you could tag each of the atoms in your body and follow them backward in time, through the air that you breathed during your life, through the food that you ate, back through the geological history of the Earth, through the ancient seas and soil, back to the formation of the Earth out of the solar nebular cloud, and then out into interstellar space, you could trace each of your atoms, those exact atoms, to *particular* massive stars in the past of our galaxy. At the end of their lifetimes, those stars exploded and spewed out their newly forged atoms into space, later to condense into planets and oceans and plants and your body at this moment. We have seen such stellar explosions with our telescopes and know they occur.

If, instead of going backward in time, let me go forward in time, to my death and beyond. The atoms in my body will remain, only they will be scattered about. Those atoms will not know where they came from, but they will have been mine. Some of them will once have been part of the memory of my mother dancing the bossa nova. Some will once have been part of the memory of the vinegary smell of my first apartment. Some will once have been part of my hand. If I could label each of my atoms at this moment, imprint each with my Social Security number, someone could follow them for the next thousand years as they floated in air, mixed with the soil, became parts of particular plants and trees, dissolved in the ocean, and then floated again to the air. And some will undoubtedly become parts of other people, particular people. So, we are literally connected to the stars, and we are literally connected to future generations of people. In this way, even in a material universe, we are connected to all things future and past.

Acknowledgments

I am grateful to Rebecca Goldstein and to Christian Mandl for helpful comments and insights on the manuscript. I thank Christof Koch and Cynthia Frantz for their conversations with me. I thank my literary agent, Deborah Schneider, for her constant enthusiasm for my work. I thank my new editor at Pantheon, Edward Kastenmeier, for his excellent guidance and editorial suggestions.

Finally, I want to pay tribute to my longtime former editor at Pantheon, Dan Frank, who left this world, and all of us, far too soon.

Notes

Introduction

9 "Does God Exist?": Alan Lightman, "Does God Exist?," *Salon,* October 2, 2011. See the rebuttal and accusation of my being an apologist for religion in "When Atheists Fib to Protect God," by Daniel Dennett, *Salon,* October 11, 2011.

9 My aim is not to prove: Recently, I had a moderated dialogue with the distinguished Islamic scholar Osman Bakar at the International Big Think Summit in Malaysia, October 10, 2021. Professor Osman strongly disagreed with me that we cannot prove the existence of God, stating that "revelation," in both the sacred books and in personal experience, shows that we know God exists.

1. The *Ka* and the *Ba*: A Brief History of the Soul, the Nonmaterial, and the Mind-Body Duality

13 This particular painting: *Lavater and Lessing Visit Moses Mendelssohn* (1856) is by Moritz D. Oppenheim.

14 "a companionable, brilliant soul": Israel Abrahams, "Men-

delssohn, Moses," *Encyclopaedia Britannica,* 11th ed. (Cambridge, UK: University of Cambridge, 1911).

14 Mendelssohn is the most famous: An excellent biography of Mendelssohn, but lacking a full set of references and sources, is Shmuel Feiner, *Moses Mendelssohn, Sage of Modernity,* trans. from the Hebrew by Anthony Berris (New Haven: Yale University Press, 2010).

15 In his salon: Ibid., p. 200.

16 "I . . . tried to adapt": Moses Mendelssohn, *Phädon, or the Immortality of the Soul,* trans. from the German by Patricia Noble (New York: Peter Lang Publishing, 2007), p. 42.

16 "There is, therefore": Ibid., p. 120.

17 "the German Socrates": Feiner, *Moses Mendelssohn,* p. 77.

18 "Liszt has many fingers": "A Heavyweight Musical Boxing Match: Franz Liszt vs. Felix Mendelssohn," *Interlude,* March 3, 2020, https://interlude.hk/a-heavyweight-musical-boxing-match-franz-liszt-vs-felix-mendelssohn/.

18 "[God] does as few": Mendelssohn, *Phädon,* p. 18.

20 "Ho, Unis!": *The Ancient Egyptian Pyramid Texts,* 2nd ed., trans. James P. Allen (Atlanta: SBL Press, 2015), p. 34.

21 "The first level thinks": Mendelssohn, *Phädon,* p. 123.

22 "As long as we trudge": Ibid., p. 83.

22 in Chinese philosophy: For the Chinese philosophy about the soul, see, for example, https://www.encyclopedia.com/environment/encyclopedias-almanacs-transcripts-and-maps/soul-chinese-concepts.

24 "The spirit soul": Srimad-Bhagavatam 7.2.22, trans. A. C. Bhaktivedanta Swami Prabhupada, https://prabhupada.io/books/sb/7/2/22.

24 "inner space": The Dalai Lama discusses the Buddhist "inner space" in the public television program *Infinite Poten-*

tial: The Life and Ideas of David Bohm (2020) directed by Paul Howard.

25 according to the Pew Research Center: https://www.pewresearch.org/fact-tank/2015/11/10/most-americans-believe-in-heaven-and-hell/.

25 YouGov: https://d25d2506sfb94s.cloudfront.net/cumulus_uploads/document/wo6pg9rb3c/Results%20for%20YouGov%20RealTime%20(Halloween%20Paranormal)%20237%2010.1.2019.xlsx%20%20[Group].pdf.

29 "the seen is the changing": Plato, *Phaedo,* trans. Benjamin Jowett, in *Great Books of the Western World,* vol. 7 (Chicago: Encyclopaedia Britannica, 1952), p. 231.

30 "occupies a larger place": Saint Augustine, letter 166.2.4, in *Letters 156–210,* trans. R. Teske (New York: New York City Press, 2004).

31 "The soul . . . seems to me": Augustine, *Greatness of the Soul* 13.22, trans. Joseph M. Colleran, in *The Greatness of the Soul, The Teacher,* ed. Johannes Quasten and Joseph Plumpe (New York: The Newman Press, 1950).

32 "The soul is defined": St. Thomas Aquinas, *Summa Theologica,* "Treatise on Man," question LXXV, trans. Fathers of the English Dominican Province, in *Great Books of the Western World,* vol. 19 (Chicago: Encyclopaedia Britannica, 1952), pp. 378–79.

33 "Some powers belong": Ibid., question LXXVII, p. 406.

34 "From that I knew": René Descartes, *Discourse on the Method of Rightly Conducting the Reason and Seeking for Truth in the Sciences* (1637), trans. from the Latin by Elizabeth S. Haldane and G. R. T. Ross, in *Great Books of the Western World,* vol. 31 (Chicago: Encyclopaedia Britannica, 1952), pp. 51–52.

35 "One cannot in any way": René Descartes, *The Passions of the Soul,* part 1, art. 30, trans. Stephen Voss (Indianapolis: Hackett Publishing Company, 1989), p. 35.

36 "Hypotheses Relating": J. C. Eccles, "Hypotheses Relating to the Brain-Mind Problem," *Nature,* 168, July 14, 1951 (4263): 53–57.

37 "the best of all possible worlds": In Leibniz's *Theodicy* (1711). See Gottfried Leibniz, *Discourse on Metaphysics and Other Essays*, trans. and ed. Daniel Garber and Roger Ariew (Indianapolis: Hackett, 1991), pp. 3–55.

37 "the true atoms of nature": Gottfried Leibniz, *Monadology,* trans. Lloyd Strickland (Edinburgh: Edinburgh University Press, 2014), axiom 3.

38 "Heaven, more than a place": https://www.youtube.com/watch?v=rWeFuPnVRGw.

38 "we are not bodies": Rabbi Micah Greenstein, interview with AL, January 5, 2016.

39 "We are certain": Mendelssohn, *Phädon,* p. 83.

39 "the contemplation of the structure": Feiner, *Moses Mendelssohn,* pp. 31–32.

40 "The passion of surprise and wonder": David Hume, "Of Miracles," in *An Enquiry Concerning Human Understanding* (1748), in *Harvard Classics,* vol. 37 (Cambridge, MA: Harvard University Press: 1910), p. 403.

40 *Wonders and the Order:* Lorraine Daston and Katharine Park, *Wonders and the Order of Nature* (New York: MIT Press/Zone Books, 1998).

42 "eternal chaotic inflation": For a review of Linde's chaotic eternal inflation cosmological model, see Andrei Linde, "The Self-Reproducing Inflationary Universe," *Scientific American,* November 1994.

2. *Primordia Rerum:* A Brief History of Materialism

44 A floor mosaic: See https://www.thecollector.com/death-in-ancient-rome/.

45 "People in good health": Thucydides, *History of the Peloponnesian War,* 2.49, in *Great Books of the Western World,* vol. 6 (Chicago: Encyclopaedia Britannica, 1952).

46 "is marked with the whip": From *Gorgias,* one of the dialogues of Plato, trans. Benjamin Jowett, in *Great Books of the Western World,* vol. 7 (Chicago: Encyclopaedia Britannica, 1952), pp. 292–93.

46 "the sea of night": Virgil, *The Aeneid,* book VI, trans. John Dryden, Project Gutenberg, 1995, https://www.gutenberg.org/files/228/228-h/228-h.htm.

47 "mist and smoke disperse": Lucretius, *De rerum natura* (*On the Nature of Things*), book III, 435–40, trans. W. H. D. Rouse (Cambridge, MA: Harvard University Press, 1982), p. 221; 830, p. 253.

47 "full of inspired brilliance": Cicero, *Epistulae ad quintum fratrem* 2.10.3, February 54 BC in *Letters to Friends,* Volume I: Letters 1–113, ed. and trans. D. R. Shackleton Bailey. Loeb Classical Library (Cambridge, MA: Harvard University Press, 2001), p. 205. Cicero's full sentence to his brother Quintus was: *Lucreti poemata, ut scribis, ita sunt, multis luminibus ingeni, multae tamen artis.*

48 a fifteenth-century Italian scholar: Poggio Bracciolini and the story of how he rescued *De rerum natura* is described in the wonderful book *The Swerve: How the World Became Modern* by Stephen Greenblatt (New York: W. W. Norton, 2011).

49 "when clouds by their collision": Lucretius, *De rerum natura,* book VI, 160–62, p. 505.

50 "The poet Titus Lucretius was born": Rudolf Helm, *Werke,* Band 7, *Die Chronik des Hieronymus/Hieronymi Chronicon* (Berlin, Boston: De Gruyter, 2013), p. 149. Also *Lucretius,* trans. W. H. D. Rouse and M. F. Smith (Cambridge, MA: Harvard University Press, 1982), p. x.

50 "delicate and composed": Lucretius, *De rerum natura,* book III, 425–40, pp. 221–23.

51 "wrecked with the mighty": Ibid., 451–52, p. 223.

52 A recent Pew survey: https://www.pewresearch.org/fact

-tank/2015/11/10/most-americans-believe-in-heaven-and-hell/.

53 "the souls of the dead are dissolved": Wang Ch'ung, *Lun Heng,* part 1, trans. from the Chinese by Alfred Forke (London: Luzac and Co., 1907), p. 207; "From the time," p. 193.

56 Al-Haytham declared: *Alhacen's Theory of Visual Perception,* book I, 6.54, vol. 2, trans. A. Mark Smith (Philadelphia: American Philosophical Society, 2001), p. 372.

57 The influential French physician: For more information on Barthez, see his entry in the *Dictionary of Scientific Biography,* vol. 1 (New York: Charles Scribner's Sons, 1981), p. 478.

58 "In living nature": Jöns Jacob Berzelius, *Lärbok i kemien* (1808), trans. and quoted in Henry M. Leicester, "Berzelius," *Dictionary of Scientific Biography,* vol. 2 (New York: Charles Scribner's Sons, 1981), p. 96.

58 "chemistry, in its application": Jean Antoine Chaptal, *Chemistry Applied to Arts and Manufactures,* vol. 1, trans. W. Nicholson (London: Richard Phillips, 1807), p. 50.

59 *This to attain:* John Milton, *Paradise Lost* (1658–63), book VIII, in *Harvard Classics,* vol. 4, ed. Charles W. Eliot (New York: P. F. Collier & Son, 1937), p. 245.

60 "in physics one must": Trans. by Jacques Roger in *Dictionary of Scientific Biography,* vol. 2 (New York: Charles Scribner's Sons, 1981), pp. 577, 579.

62 any solid object: "On Floating Bodies" (ca. 250 BC), in *The Works of Archimedes,* ed. T. L. Heath (Cambridge: Cambridge University Press, 1897), book I, prop. 5.

63 law of falling bodies: For Galileo's law of falling bodies, see *Dialogues Concerning the Two New Sciences,* third day, theorem II, prop. 2, trans. from the Italian by Henry Crew and Alfonso de Salvio, in *Great Books of the Western World,* vol. 28 (Chicago: Encyclopaedia Britannica, 1952), p. 206.

65 "Anyone will then understand": Galileo Galilei, *Sidereus nuncius* (1610) in its original Latin, trans. and with notes by Albert Van Helden (Chicago: University of Chicago Press, 1989), p. 36.

66 "I wait to hear spoutings": Galileo, *Opere,* 11, no. 675, 295–97, p. 296, trans. from the Italian in John Michael Lewis, *Galileo in France: French Reactions to the Theories and Trial of Galileo* (New York: Peter Lang Publishing, 2006) p. 94.

66 That divine substance: Aristotle, *On the Heavens,* book I, ch. 3, trans. from the Greek by W. K. C. Guthrie, in the *Loeb Classical Library,* vol. 6 (Cambridge, MA: Harvard University Press, 1971), pp. 23–25.

67 "such an elegant": Richard Feynman, *The Character of Physical Law* (Cambridge: MIT Press, 1965), p. 14.

71 "Forces are causes": *Annalen der Chemie und Pharmacie* 42 (1843), trans. from the French by G. C. Foster in *Philosophical Magazine,* series 4, vol. 24 (1862), p. 271; reprinted in *A Source Book in Physics,* ed. W. F. Magie (New York: McGraw-Hill, 1935), pp. 197–201.

73 "If men saw that a limit": Lucretius, *De rerum natura,* book I, 107–110, p. 13.

74 "at times quite uncertain": Ibid., book II, 216–60, pp. 113–15.

74 "When abundant matter": Ibid., book II, 1067–176, p. 179.

75 "I do not grieve": Plato, *Phaedo,* trans. Benjamin Jowett, in *Harvard Classics,* vol. 2 (New York: P. F. Collier & Son, 1909), p. 51.

75 "As, therefore, all men [desire]": Saint Augustine, *On the Trinity,* book XIII, ch. 8, trans. Arthur West Haddan (Edmond, OK: Veritas Splendor Publications, 2012), http://www.logoslibrary.org/augustine/trinity/1308.html.

77 And a relatively recent survey: Of two thousand practicing physicians, "Survey Shows That Physicians Are

More Religious Than Expected," University of Chicago Medicine, June 22, 2005, https://www.uchicagomedicine.org/forefront/news/survey-shows-that-physicians-are-more-religious-than-expected.

79 "I hope that": Interview with AL, Massachusetts General Hospital, July 17, 2019.

80 Even today, according to: "Religion Among the Millennials," Pew Research Center, https://www.pewforum.org/2010/02/17/religion-among-the-millennials/.

80 "We want miracles": Interview with AL, Cambridge, MA, October 14, 2020.

82 "It is your merit,": Lucretius, *De rerum natura,* book I, 140–45, p. 15.

82 "So trivial": Ibid., book III, 320–22, p. 213.

82 "Golden images of youths": Ibid., book II, 25–35, p. 97.

82 "As soon as the brightness": Ibid., book IV, 209–15, p. 293.

3. Neurons and I: The Emergence of Consciousness from the Material Brain

83 "I've wondered about dogs": Steve Paulson, "What is this thing called consciousness?" *Nautilus,* April 6, 2017.

83 "It's much more widespread": Kevin Berger, "Ingenious: Christof Koch," *Nautilus,* April 15, 2019.

86 "Fundamentally an organism": Thomas Nagel, "What Is It Like to Be a Bat?" *The Philosophical Review* 83, no. 4 (October 1974): 435–50.

86 "Nothing we can think": A. Revonsuo, *Consciousness: The Science of Subjectivity* (Hove: Psychology Press, 2010), p. 30.

87 "What we have": Peter McGinn, *The Mysterious Flame, Conscious Minds in a Material World* (New York: Basic Books, 1999), p. 212.

87 "our intelligence is wrongly": Ibid., p. xi.

88 His 2004 book: Christof Koch, *The Quest for Consciousness* (Englewood, CO: Roberts and Company, 2004).

88 "continuous flash suppression": N. Tsuchiya and C. Koch, "Continuous Flash Suppression Reduces Negative After Images," *Nature Neuroscience* 8, no. 8 (2005): 1096–101.

89 In 1990, Koch and Francis Crick: F. C. Crick and C. Koch, "Towards a neurobiological theory of consciousness," *Seminars in Neuroscience* 2, no. 263 (1990). For earlier work by von der Malsburg in the 1980s, a review is given in C. von der Malsburg, "The what and why of binding: The modeler's perspective," *Neuron* 24, no. 95 (1999).

90 The attention proposal: The research findings of Desimone and Baldauf were published in "Neural Mechanisms for Object-Based Attention," *Science* 344, no. 6182 (April 2014): 424–27.

92 "As we learn more": This and other quotes from Desimone come from my interviewing him in his office at MIT on September 17, 2014.

93 "Life itself is such a mystery": My Zoom interview with Koch on July 15, 2021.

93 "So how do I know": Ibid.

95 The most important measure of intelligence: For a discussion of brain weights and intelligence, see Christof Koch, "Does Brain Size Matter?" *Scientific American Mind,* January/February 2016, p. 22.

99 "neuronal correlates of consciousness": For one of the earliest papers, see Crick and Koch, "Towards a Neurobiological Theory of Consciousness." For one of the more recent papers, see Todd E. Feinberg and Jon Mallatt, "Phenomenal Consciousness and Emergence: Eliminating the Explanatory Gap," *Frontiers of Psychology,* June 12, 2020.

100 In 1970, the British American neuroscientist: J. O'Keefe and J. Dostrovsky, "The hippocampus as a spatial map. Preliminary evidence from unit activity in the freely-moving rat," *Brain Research* 34 (1971): 171–75.

100 Edvard Moser and May-Britt Moser: T. Hafting, M. Fyhn, S. Molden, et al., "Microstructure of a spatial map in the entorhinal cortex," *Nature* 436, no. 7052 (2005): 801–6.

101 It has long been known: See, for example, Matthias Stangl et al., "Compromised Grid-Cell-like Representations in Old Age as a Key Mechanism to Explain Age-Related Navigational Deficits," *Current Biology* 28, no. 7 (April 2, 2018): 1108–115.

101 "Understanding the material basis": Koch, *The Quest for Consciousness,* p. 10.

102 "How well does the patient": The Awareness Questionnaire, The Center for Outcome Measurement in Brain Injury, 2004, http://www.tbims.org/combi/aq.

103 Not surprisingly, the results: See, for example, Mark Sherer, Tess Hart, Todd Nick, et al., "Early Impaired Self-Awareness After Traumatic Brain Injury," *Archives of Physical Medical Rehabilitation* 84, no. 2 (February 2003): 168–76.

103 Autobiographical memory: For a review of memory loss after brain injury, see Eli Vakil, "The Effect of Moderate to Severe Traumatic Brain Injury (TBI) on Different Aspects of Memory: A Selective Review," *Journal of Clinical and Experimental Neuropsychology* 27, no. 8 (2005): 977–1021.

104 "When leaving my doctor's surgery": https://www.dementia.org.au/about-us/news-and-stories/stories/day-25-leo-tas. His story at the now obsolete site: "In Our Own Words: Younger Onset Dementia," https://fightdementia.org.au/files/20101027-Nat-YOD-InOurOwnWords.pdf.

105 LSD changes the way neurons communicate: For studies of the effect of LSD on serotonin, see, for example, Wacker et al., "Crystal Structure of an LSD-Bound Human Serotonin Receptor," *Cell* 168 (January 26, 2017): 377–89.

105 "All of a sudden there was": https://www.reddit.com/r/LSD/comments/4i3c70/diary_of_a_solo_acid_trip progression_and/.

107 Dolphins, which have almost: Diana Reiss and Lori Marino, "Mirror self-recognition in the bottlenose dolphin: A case of cognitive convergence," *Publications of the National Academy of Sciences,* 98, no. 10 (May 8, 2001): 5937–42.

107 A hilarious video on YouTube: https://www.youtube.com/watch?v=YbdNtC4V3IM.

108 A video posted in February 2021: https://www.youtube.com/watch?v=UrONJIoaIgU.

108 Neuroscientist and psychologist Michael Beran: M. J. Beran, J. D. Smith, and B. M. Perdue, "Language-trained chimpanzees name what they have seen, but look first at what they have not seen," *Psychological Science* 24, no. 5 (May 2013): 660–66.

109 James R. Anderson and colleagues: James R. Anderson, Alasdair Gillies, and Louise C. Lock, "Pan Thanatology," *Current Biology* 20, no. 8 (April 27, 2010): PR349–51.

110 Psychiatrist and neurologist Todd Feinberg: For work by Feinberg and Mallatt on consciousness, see Todd E. Feinberg and Jon Mallatt, "Phenomenal Consciousness and Emergence: Eliminating the Explanatory Gap," *Frontiers of Psychology,* June 12, 2020.

112 "integrated information theory": G. Tononi, "An information integration theory of consciousness," *BMC Neuroscience* 5, no. 42 (2004); G. Tononi and C. Koch, "Consciousness: Here, there and everywhere?" *Philosoph-*

ical Transactions of the Royal Society B 370: 20140167 (2015).

112 "If you build a neuromorphic brain": Berger, "Ingenious: Christof Koch."

114 "There is a two-way disassociation": My Zoom conversation with Koch, July 15, 2021.

115 The modern understanding of emergentism: For Mill's discussion of emergentism, see John Stewart Mill, *A System of Logic, Ratiocinative, and Inductive* (London: Longmans, Green and Co., 1843); 8th ed. (New York: Harper and Brothers, 1882), ch. 6, p. 459.

116 the human brain can store: Storage capacity of the largest computer: https://www.forbes.com/sites/aarontilley/2017/05/16/hpe-160-terabytes-memory/?sh=62c847b6383f; storage capacity of the human brain: https://www.scientificamerican.com/article/what-is-the-memory-capacity/.

116 "The abilities of coalitions": Koch, *The Quest for Consciousness,* p. 10.

120 "First, it is a true experience": My Zoom interview with Koch on July 15, 2021.

4. To See a World in a Grain of Sand: From Consciousness to Spirituality

121 "To See a World in a Grain of Sand": "To See a World in a Grain of Sand" (1803) is the first line of "Auguries of Innocence," by William Blake.

122 "I remember the night": William James, *Varieties of Religious Experience* (1902), BiblioBazaar ed. (2007), p. 71.

123 "Thou [God] hast made me endless": Rabindranath Tagore, *Gitanjali,* trans. W. B. Yeats (New York: The MacMillan Company, 1916), stanza 1, p. 1; "the same stream of life," stanza 69, pp. 64–65.

123 "When I was midway on the mountain": Ibn Ishaq, *The Life of Muhammad,* trans. Alfred Guillaume (Oxford: Oxford University Press, 1967).

124 "And the angel of the LORD": Exodus 3:2.

125 evolutionary biologists Stephen Jay Gould: See Stephen Jay Gould and Richard Lewontin, "The Spandrels of San Marco and the Panglossian Paradigm: A Critique of the Adaptationist Programme," *Proceedings of the Royal Society of London B,* 205, no. 1161 (1979).

127 "So poor is nature": Ralph Waldo Emerson, "Nature" (1836), in *The Harvard Classics,* vol. 5 (New York: P. F. Collier & Son, 1909), p. 229.

128 "the innate tendency": E. O. Wilson, *Biophilia* (Cambridge, MA: Harvard University Press, 1984), prologue.

128 The term was first coined: Erich Fromm, *The Heart of Man* (New York: Harper and Row, 1964).

128 "the crucial first step": Wilson, *Biophilia,* pp. 105–6.

129 tropical fish called guppies produce: For the work of David Reznick and colleagues on guppies, see, for example, Ronald D. Bassar, Michael C. Marshall, Andrés López-Sepulcre, et al., "Local adaptation in Trinidadian guppies alters ecosystem processes," *Proceedings of the National Academy of Sciences* 107, no. 8 (February 23, 2010): 3616.

130 "Connectedness to Nature Scale": F. S. Mayer and C. M. Frantz, "The connectedness to nature scale: A measure of individuals' feeling in community with nature," *Journal of Environmental Psychology* 24 (2004): 503–15.

131 well-developed methods of measuring happiness: For a review of different methods for measuring happiness and well-being, see Philip J. Cooke, Timothy P. Melchert, and Korey Connor, "Measuring Well Being: A Review of Instruments," *The Counseling Psychologist* 44, no. 5 (July 1, 2016): 730–57.

131 psychologist Colin Capaldi: Colin Capaldi, Raelyne L. Dopko, and John Michael Zelenski, "The relationship between nature connectedness and happiness: A meta-analysis," *Frontiers in Psychology* 5 (September 2014).

132 "If we are more in tune": AL interview with Cindy Frantz, August 11, 2021. All quotes from Frantz come from this interview.

133 "cooperation is directed": Stuart West, "Competition Between Groups Drives Cooperation within Groups," the Leakey Foundation, August 1, 2016, https://leakeyfoundation.org/the-evolutionary-benefits-of-cooperation/.

134 "One of the most challenging": AL interview with Nicholson Browning, January 27–31, 2021.

135 "UCLA Loneliness Scale": D. Russell, L. A. Peplau, and C. E. Cutrona, "The revised UCLA loneliness scale: Concurrent and discriminant validity evidence," *Journal of Personality and Social Psychology* 39 (1980): 472–80.

135 In a study of 240 people: Andrew Steptoe, Natalie Owen Sabine, R. Kunz-Ebrecht, and Lena Brydon, "Loneliness and neuroendocrine, cardiovascular, and inflammatory stress responses in middle-aged men and women," *Psychoneuroendocrinology* 29, no. 5 (June 2004): 593–611.

136 "borrowing the pain signal": C. DeWall, T. Deckman, R. S. Pond, and I. Bonser, "Belongingness as a Core Personality Trait: How Social Exclusion Influences Social Functioning and Personality Expression," *Journal of Personality* 79, no. 6 (2011): 979–1012. See also J. Panksepp, B. H. Herman, R. Conner, et al., "The biology of social attachments: Opiates alleviate separation distress," *Biological Psychiatry* 13 (1978): 607.

136 Geoff MacDonald: See G. MacDonald and M. R. Leary, "Why does social exclusion hurt? The relationship between social and physical pain," *Psychological Bulletin* 131 (2005): 202–23.

137 American psychologist Harry Harlow: See: H. F. Harlow, "The nature of love," *American Psychologist* 13, no. 12 (1958): 673–85; H. F. Harlow, R. O. Dodsworth, and M. K. Harlow, "Total Social Isolation in Monkeys," *Pro-*

ceedings of the National Academy of Sciences 54, no. 1 (June 1965): 90–97; H. A. Cross and H. F. Harlow, "Prolonged and progressive effects of partial isolation on the behavior of macaque monkeys," *Journal of Experimental Research in Personality* 1 (1965): 39–49.

137 pediatrician Ruth Feldman: Ruth Feldman, Arthur I. Eidelman, Lea Sirota, and Aron Weller, "Comparison of skin-to-skin (kangaroo) and traditional care: Parenting outcomes and preterm infant development," *Pediatrics* 110, 1 pt. 1 (July 2002): 16–26.

138 "There is no 'I' in nature": Cynthia Frantz, F. Stephan Mayer, Chelsey Norton, and Mindi Rock, "There is no 'I' in nature: The influence of self-awareness on connectedness to nature," *Journal of Environmental Psychology* 25 (2005): 427–36.

139 substantial differences between Westerners: A good analysis of the differences between Western and Eastern psychology can be found in *The Weirdest People in the World* (New York: Farrar, Straus, and Giroux, 2020) by Joseph Henrich, a leading researcher in this area. See also the article "How East and West Think in Profoundly Different Ways," by David Robson, *The Human Planet,* January 19, 2017.

140 "The unexamined life": Plato, *Apology,* 38, trans. Benjamin Jowett, in *Great Books of the Western World,* vol. 7 (Chicago: Encyclopaedia Britannica, 1952), p. 210.

140 "In planning for others": *The Analects of Confucius,* 1.4, trans. Robert Eno, https://chinatxt.sitehost.iu.edu/Analects_of_Confucius_(Eno-2015).pdf.

140 "The Significance of the Frontier": Frederick Jackson Turner, "The Significance of the Frontier in American History" (1893), https://www.historians.org/about-aha-and-membership/aha-history-and-archives/historical-archives/the-significance-of-the-frontier-in-american-history-(1893).

141 social psychologist Shinobu Kitayama: See Shinobu Kitayama, Keiko Ishii, Toshie Imada, et al., "Voluntary settlement and the spirit of independence: Evidence from Japan's northern frontier," *Journal of Personality and Social Psychology* 91, no. 3 (2006): 369.

142 "How strange is the lot of us mortals": Originally published in *Forum and Century* 84 (1931): 193–94; reprinted in Albert Einstein, *Ideas and Opinions,* trans. Sonja Bargmann (New York: The Modern Library, 1994), p. 8.

143 "And having made [the universe]": Plato, *Timaeus,* trans. Benjamin Jowett, in *Great Books of the Western World,* vol. 7 (Chicago: University of Chicago Press, 1952), p. 452.

143 "oceanic feeling": Sigmund Freud, *Civilization and Its Discontents* (1930), trans. James Strachey (New York: W. W. Norton, 1961), pp. 11–12.

143 "the idea of death": Ernest Becker, *The Denial of Death* (New York: The Free Press, 1973), p. xvii.

145 "When we look back": AL interview with Kip Thorne, August 16, 2021.

146 "cosmic biocentrism": Alan Lightman, *Probable Impossibilities* (New York: Pantheon Books, 2021), p. 162.

147 "man's love for natural colors": H. Iltis, "Why man needs open space: The basic optimum human environment," in *The Urban Setting Symposium,* ed. S. H. Taylor (New London, CT: Connecticut College, 1980), p. 3.

148 "When we behold a male bird": Charles Darwin, *The Descent of Man* (1871), ch. 3, "Sense of Beauty," in *Great Books of the Western World,* vol. 49 (Chicago: University of Chicago Press, 1952), p. 301.

148 "Psychoanalysis has less to say": Sigmund Freud, *Civilization and Its Discontents* (1929), ch. 2, in *Great Books of the Western World,* vol. 54 (Chicago: University of Chicago Press, 1952), p. 775.

149 Take the golden ratio: Mathematically, if a is the larger quantity and b the smaller, then, a/b is a golden ratio if $a/b = (a + b)/a$. Dividing the numerator and denominator on the right-hand side by b, we get $a/b = (a/b + 1) / a/b$. We can solve this quadratic equation for a/b, the golden ratio, getting: $a/b = (1+ \sqrt{5})/2$.

152 mechanical engineer Adrian Bejan: See Adrian Bejan, "The golden ratio predicted: Vision, cognition and locomotion as a single design in nature" *International Journal of Design and Nature and Ecodynamics* 4, no. 2 (2009): 97–104.

154 "Approaching awe": Dacher Keltner and Jonathan Haidt, "Approaching awe, a moral, spiritual, and aesthetic emotion," *Cognition and Emotion* 17, no. 2 (2003): 297–314.

155 "Up and up, the Incense-burner": http://www.mountainsongs.net/poem_.php?id=904.

156 "Suppose you have something": Alan Lightman and Roberta Brawer, *Origins: The Lives and Worlds of Modern Cosmologists* (Cambridge, MA: Harvard University Press, 1990), pp. 433–34.

158 The creative transcendent would occur: Graham Wallas, *The Art of Thought* (London: C. A. Watts and Company, 1926).

159 "Every day I seated myself": Henri Poincaré, *The Foundations of Science,* trans. George Bruce Halsted (New York: The Science Press, 1913), p. 387.

160 "All my awareness": AL interview with Paul Ingbretson, August 26, 2021.

161 "I made straight for Heligoland": Werner Heisenberg, *Physics and Beyond,* trans. from the German by Arnold J. Pomerans (New York: Harper and Row, 1971), pp. 60–61.

162 "A novelist's chief desire": Virginia Woolf, "Professions for a Woman" (1931), lecture delivered before a branch of the National Society for Women's Service on January 21,

1931, published posthumously in *The Death of the Moth and Other Essays*. See, for example, (Victoria BC, Canada: Rare Treasures Press, 2000), p. 2017.

5. My Atoms and Yours: Science and Spirituality in the World of Today

168 In September 2018, I debated: https://www.youtube.com/watch?v=eSCDfjTDVCk.

168 Dawkins dismisses people of faith: Dawkins's comments about religion and faith, in a speech at the Edinburgh International Science Festival, in 1992, quoted in "A scientist's case against God," *The Independent* (London), April 20, 1992, p. 17: "Faith is the great cop-out, the great excuse to evade the need to think and evaluate evidence." In an article titled "Has the World Changed? Part Two" in *The Guardian,* October 11, 2001, Dawkins wrote, "Many of us saw religion as harmless nonsense. Beliefs might lack all supporting evidence but, we thought, if people needed a crutch for consolation, where's the harm?"

169 "The most beautiful experience": Originally published in *Forum and Century* 84 (1931): 193–94; reprinted in Albert Einstein, *Ideas and Opinions,* trans. Sonja Barmann (New York: The Modern Library, 1994), p. 11.

Illustration Credits

Page

15 Moses Mendelssohn painting, after Anton Graff, Wikimedia Commons

23 Chinese seal script for *hun,* Wikimedia Commons

37 Gottfried Wilhelm Leibniz, by Andreas Scheits (1655–1735), Wikimedia Commons

46 Marble carving of punishment of Ixion, image ID: HJ3883, Chris Hellier/Alamy Stock Photo, with permission

50 Lucretius, unknown author, Wikimedia Commons

56 Al-Haytham diagram from *Alhacen's Theory of Visual Perception* (*De aspectibus*), vol. 2, trans. A. Mark Smith (Philadelphia: American Philosophical Society, 2001), p. 406

64 Galileo's telescope. Credit: Science Museum Group, with permission

65 Galileo's drawing of the Moon. Galileo Galilei, *Sidereus nuncius,* trans. Albert Van Helden (Chicago: University of Chicago Press, 1989), p. 44

66 Cover of Newton's *Principia*. Isaac Newton, *Principia* (1729), trans. Andrew Motte, rev. Florian Cajori (Berkeley: University of California Press, 1934), p. xii

68 Electromagnetic wave, Wikimedia Commons, P.wormer, CC BY-SA 3.0

89 Christof Koch, © Fatma Imamoglu, with permission

97 Clinton neuron, from *The Quest for Consciousness* by Christof Koch (Englewood, CO: Roberts and Company Publishers, 2004), p. 30, modified from Kreiman, "On the Neuronal Activity of the Human Brain During Visual

Recognition, Imagery and Binocular Rivalry," PhD thesis, Caltech, 2001, with permission

117 STAT3 protein, photo by Trcum, Wikimedia Commons

117 Snowflake designs: photo by Wilson Bentley, Wikimedia Commons

118 Termite cathedral: photo by J. Brew, Wikimedia Commons

131 Cindy Frantz, photo by Tanya Rosen Jones, with permission

133 Girls together, photo by Jodi Hilton

150 Fibonacci spiral, by Jahorbr, Wikimedia Commons

150 Seashell, by Chris 73, Wikimedia Commons

151 *Aloe polyphylla*, photo by brewbooks, Wikimedia Commons

151 Great Pyramid of Giza, photo by Nina, Wikimedia Commons

152 Right triangle, drawn by the author

152 CN Tower in Toronto, photo by Diego Delso, Wikimedia Commons

A Note About the Author

Alan Lightman—who worked for many years as a theoretical physicist—is the author of seven novels, including the international best seller *Einstein's Dreams,* as well as *The Diagnosis,* a finalist for the National Book Award. He is also the author of a memoir, three collections of essays, and several books on science. His work has appeared in *The Atlantic, Granta, Harper's Magazine, The New Yorker, The New York Review of Books, Salon,* and *Nature,* among other publications. He has taught at Harvard and at MIT, where he was the first person to receive a dual faculty appointment in science and the humanities. He is currently professor of the practice of the humanities at MIT. He lives in the Boston area.

A Note on the Type

The text of this book was set in Imprint, a typeface originally produced in 1913 by the Monotype Corporation for the renowned printer Gerard Meynell's short-lived periodical *The Imprint*. A collaboration between Meynell, Edward Johnston, and J. H. Mason, the face is modeled on Caslon, but has a larger x-height and a greatly regularized italic. Imprint has the distinction of being the first original typeface designed specifically for machine composition.

Composed by North Market Street Graphics,
Lancaster, Pennsylvania

Printed and bound by Berryville Graphics,
Berryville, Virginia

Designed by Betty Lew